Printed in the United States
By Bookmasters

T0214327

Lecture Notes in Computer Science   11758

More information about this series at http://www.springer.com/series/7407

Brijesh Dongol · Luigia Petre ·
Graeme Smith (Eds.)

Formal Methods
Teaching

Third International Workshop and Tutorial, FMTea 2019
Held as Part of the Third World Congress on Formal Methods, FM 2019
Porto, Portugal, October 7, 2019
Proceedings

 Springer

Editors
Brijesh Dongol 🄳
University of Surrey
Guildford, UK

Luigia Petre
Åbo Akademi University
Turku, Finland

Graeme Smith
The University of Queensland
Brisbane, QLD, Australia

ISSN 0302-9743   ISSN 1611-3349   (electronic)
Lecture Notes in Computer Science
ISBN 978-3-030-32440-7   ISBN 978-3-030-32441-4   (eBook)
https://doi.org/10.1007/978-3-030-32441-4

LNCS Sublibrary: SL1 – Theoretical Computer Science and General Issues

This Springer imprint is published by the registered company Springer Nature Switzerland AG
The registered company address is: Gewerbestrasse 11, 6330 Cham, Switzerland

Preface

Formal methods provide software engineers with tools and techniques for rigorously reasoning about the correctness of systems. While formal methods are being increasingly used in industry, university curricula are not adapting at the same pace. Some existing formal methods courses interest and challenge students, whereas others fail to ignite student motivation. We need to find ways to teach formal methods to the next generation, and doing so will require us to adapt our teaching to 21st-century students.

FMTea 2019 (Formal Methods Teaching Workshop and Tutorial) was a combined workshop and tutorial on teaching formal methods held on October 7, 2019, as part of the Third World Congress on Formal Methods, FM 2019, in Porto, Portugal. The event was organized by FME's Teaching Committee whose aim is to support worldwide improvement in teaching and learning formal methods.

The workshop received 22 regular submissions, 14 of which were selected for publication after a single blind review in which all papers were reviewed by two members of the Program Committee or their delegates. Additionally, an invited submission by Tony Hoare (co-written with Alexandra Mendes and João F. Ferreira) was similarly reviewed and selected for publication bringing the total number of papers to 15.

To encourage community-building discussions on the topic of teaching formal methods, we divided the regular presentations into three panels in which authors briefly introduced their papers and then contributed to a lively discussion of their topic and its alternatives together with their peers and the audience. The first two panels were on using state-of-the-art tools for teaching program verification and program development, respectively, while the third was on novel techniques for effectively teaching formal methods to future formal methods scientists.

The technical program of FMTea 2019 also included three invited talks:

- a lecture by Carroll Morgan (University of New South Wales and Data61, Australia) on his approach to, and experiences with, teaching formal methods to undergraduate students
- a tutorial presentation by Holger Hermanns (Saarland University, Germany) on award-winning experiences with teaching concurrency theory and concurrent programming using pseuCo
- a tutorial presentation by Bas Luttik (Eindhoven University of Technology, The Netherlands) on experiences with online teaching to prepare Bachelor-level students for learning formal methods

We would like to thank all of our invited speakers for agreeing to present at our workshop and for the exciting and inspiring ideas they brought to it.

FMTea 2019 would not have been possible without the support of the FM 2019 general chair, José Nuno Oliveira, the workshop and tutorial chairs, Nelma Moreira and

Emil Sekerinski, and the numerous people involved in the local organization of FM 2019. We are grateful for their enthusiasm and dedication. We would also like to thank the Program Committee and reviewers for their expert opinions on the papers we received, and of course the authors for sharing their innovative teaching practices.

Finally, we acknowledge EasyChair, which supported us in the submission and reviewing process, as well as in generating these proceedings and the FMTea 2019 program.

August 2019   Brijesh Dongol
                                                                            Luigia Petre
                                                                         Graeme Smith

Organization

Program Committee

Brijesh Dongol	University of Surrey, UK
Catherine Dubois	ENSIIE, France
João F. Ferreira	INESC TEC & University of Lisbon, Portugal
Rustan Leino	Amazon, USA
Alexandra Mendes	INESC TEC & Universidade da Beira Interior, Portugal
Luigia Petre	Åbo Akademi University, Finland
Leila Ribeiro	Universidade Federal do Rio Grande do Sul, Brazil
Pierluigi San Pietro	Politecnico di Milano, Italy
Graeme Smith	The University of Queensland, Australia
Kenji Taguchi	CAV Technologies Co., Ltd, Japan

Additional Reviewers

Boureanu, Ioana

Invited/Tutorial Lectures

Is Formal Methods *Really* Essential?

Carroll Morgan[1,2]

[1] School of Computer Science and Engineering,
University of New South Wales, Australia
[2] Data61, Australia

Googling *Math is fun.* returns close to 300 million hits.

Is that part of "the problem" for maths education? Maybe if one says so emphatically that a topic is fun, it reveals actually that it needs special treatment, that it's optional and that students must be enticed to learn it. Of course being a full-blown mathematician is optional—but it is self-evident that being able to deploy *elementary* maths in everyday life is not optional, or at least should not be. Yet many adults cannot do even simple arithmetic.

Perhaps Formal Methods is similar: by insisting that it's essential, we might be losing some of our leverage. Of course not everyone has to be a "neat" (vs. a "scruffy"). But it is still true that an *appreciation* of algorithmic rigour should be compulsory for beginning programmers, together with some idea of how to achieve it in what has become their everyday life. Yet there are many experienced programmers who have never heard of invariants.

So maybe there's a place for *elementary formal methods* –for *FM* by stealth– to be learned at the same time as one's first-year introduction to programming: not singled out, not separated (and certainly not labelled "formal" or "elementary"), and without any extra prerequisites (like logic) beyond what is required for beginners already. Teach assertions about assignments, conditionals, loops (what they establish) at the same time, at first encounter, just as we teach already their syntax (how to write them) and their operational aspects (what they do). And bring to that as much informal intuition as we can muster: use hand-waving, pictures... and even flowcharts, where it all started.

Formal Methods? Let's not call it that: let's call it Programming.

PSEUCO.COM

Felix Freiberger[1,2] and Holger Hermanns[1,3]

[1] Saarland University, Saarland Informatics Campus, Saarbrücken, Germany
[2] Saarbrücken Graduate School of Computer Science,
Saarland Informatics Campus, Saarbrücken, Germany
[3] Institute of Intelligent Software, Guangzhou, China

Abstract. This tutorial presents PSEUCO [1], an academic programming language designed to teach concurrent programming. The language features a heavily simplified Java-like look and feel. It supports shared-memory as well as message-passing concurrent programming primitives. The behaviour of PSEUCO programs is described by a formal semantics mapping on value-passing CCS or coloured Petri nets [2], and is made executable using compilation to Java. PSEUCO is not only a language but an interactive experience: PSEUCO.COM [1] provides access to a web application designed for first hands-on experiences with CCS and with concurrent programming patterns, supported by a rich and growing toolset. It provides an environment for students to experiment with and understand the mechanics of the fundamental building blocks of concurrency theory and concurrent programming based on a complete model of the program behaviour. PSEUCO and PSEUCO.COM constitute the centerpiece of an award-winning lecture series, mandatory for Bachelor students at Saarland Informatics Campus.

Acknowledgments. This work was partially supported by the ERC Advanced Investigators Grant 695614 (POWVER) and by the Deutsche Forschungsgemeinschaft (DFG, German Research Foundation) – project number 389792660 – TRR 248 (see https://perspicuous-computing.science).

References

1. Biewer, S., Freiberger, F., Held, P.L., Hermanns, H.: Teaching academic concurrency to amazing students. In: Aceto, L., Bacci, G., Bacci, G., Ingólfsdóttir, A., Legay, A., Mardare, R. (eds.) Models, Algorithms, Logics and Tools. LNCS, vol. 10460, pp. 170–195. Springer, Cham (2017). https://doi.org/10.1007/978-3-319-63121-9_9
2. Freiberger, F., Hermanns, H.: Concurrent programming from PSEUCO to petri. In: Donatelli, S., Haar, S. (eds.) PETRI NETS 2019. LNCS, vol. 11522, pp. 279–297. Springer, Cham (2019). https://doi.org/10.1007/978-3-030-21571-2_16

Efficient Online Homologation to Prepare Students for Formal Methods Courses

Bas Luttik

Eindhoven University of Technology, The Netherlands

Abstract. At Eindhoven University of Technology, the majority of students enrolling in our master programme on Embedded Systems have one or more deficiencies in prerequisite bachelor-level knowledge and skills. In the past, we tried to determine such deficiencies on the basis of application data (e.g., a transcript of their bachelor programme), and students were then required to repair them by including one or two bachelor courses in their study programme. This approach was found to be both unreliable and inefficient. To improve, we have developed an online homologation recommendation tool by which students can determine to which extent they satisfy the prerequisites of the programme and fully automatically get a recommendation on how to repair deficiencies. Furthermore, we have developed several online self-study homologation modules.

In my talk, I will discuss my experiences with developing both the homologation recommendation tool and the online homologation module *Logic and Set Theory*, which addresses prerequisites for the mandatory formal methods course that is part of the Embedded Systems programme. The homologation module consists of over 50 short videoclips and a week-by-week exercise programme. In our experience, the material successfully and efficiently prepares master-level students for an exam of our bachelor course *Logic and Set Theory*.

Contents

Tutorial Lectures

Logic, Algebra, and Geometry at the Foundation of Computer Science

Tony Hoare[1,2], Alexandra Mendes[3,4], and João F. Ferreira[5(✉)]

[1] Microsoft Research, Cambridge, UK
[2] Cambridge University Computing Laboratory, Cambridge, UK
[3] HASLab, INESC TEC, Porto, Portugal
[4] Department of Informatics, Universidade da Beira Interior, Covilhã, Portugal
[5] INESC-ID & Instituto Superior Técnico, University of Lisbon, Lisbon, Portugal
joao@joaoff.com

Abstract. This paper shows by examples how the Theory of Programming can be taught to first-year CS undergraduates. The only prerequisite is their High School acquaintance with algebra, geometry, and propositional calculus. The main purpose of teaching the subject is to support practical programming assignments and projects throughout the degree course. The aims would be to increase the student's enjoyment of programming, reduce the workload, and increase the prospect of success.

Keywords: Algebra · Logic · Geometry · Teaching formal methods · Unifying theories of programming

1 Introduction

The Theory of Programming lies at the foundation of modern development environments for software, now widely used in industry. Computer Science graduates who understand the rationale of programming tools, and who have experience of their use, are urgently needed in industry to maintain the current rate of innovations and improvements in software products installed worldwide.

We put forward the following theses:

1. The fundamental ideas of the Theory of Programming were originally formulated by great philosophers, mathematicians, geometers and logicians, dating back to antiquity.
2. These ideas can be taught as an aid to practical programming throughout a degree course in Computer Science. The desirable initial level of Math for first-year CS students is that of High School courses in Algebra, Geometry and Propositional Logic.
3. The ideas should form the basis of a student-oriented Integrated Development Environment (IDE), needed to support students in understanding requirements, in designing solutions, in coding programs, in testing them, and in diagnosing and debugging their errors.

© Springer Nature Switzerland AG 2019
B. Dongol et al. (Eds.): FMTea 2019, LNCS 11758, pp. 3–20, 2019.
https://doi.org/10.1007/978-3-030-32441-4_1

One of the goals of this paper is to contribute to the challenge posed by Carroll Morgan in [21]:

> *Invariants, assertions and static reasoning should be as self-evidently part of the introductory Computer Science curriculum as are types, variables, control structures and I/O in the students' very first programming language.*
> *Can you help to bring that about?*

Paper Structure. In this paper we provide examples of material that can be taught to first-year CS undergraduates. In Sect. 2, we introduce the underlying concepts of algebra and logic. These are then applied to the execution of computer programs: in Sect. 3 we discuss the familiar topic of sequential composition and in Sect. 4 we move on to concurrent composition. Section 4 includes material suitable for a more advanced and elective course in formal methods delivered at later stage in the syllabus, where we show how two familiar and widely used theories of programming can be unified. After presenting in Sect. 5 some related work, we conclude in Sect. 6, where we also briefly suggest directions for future work.

2   Algebra and Logic

This section introduces the underlying concepts of algebra and logic. The first subsection is entirely elementary, but it proves some essential theorems that will be used in later sections. The second subsection shows how familiar logical proof rules can be derived from the algebra. The third subsection introduces spatial and temporal reasoning about the execution of computer programs.

2.1   Boolean Algebra

Boolean Algebra, which is widely taught at the beginning of degree courses in Mathematics and in Philosophy, is doubly relevant in a Computer Science course, both for Hardware Design and for Program Development.

Boolean Algebra is named for the nineteenth century mathematician George Boole (1815–1864). His father was a shoe-maker in Lincoln, where he attended primary school. His father died when he was aged 16, and he became the family breadwinner, working as a schoolmaster. At age 25 he was running a boarding school in Lincoln, where he was recognised as a local civic dignitary. He learnt mathematics from books lent to him by friendly mathematicians. At the age of 34, he was appointed as

George Boole
(1815–1864)

first Professor of Mathematics at the newly founded Queen's College in Cork. He published a number of articles in the humanities, and wrote several mathematical textbooks. But he is now best known for his logical investigations of the Laws of Thought [6], which he published in 1854 and where he proposed the binary algebraic operators *not, and, or,* and a binary comparison for predicates as the foundation for a deductive logic of propositions.

Disjunction. Disjunction is denoted as $\vee$ (read as 'or') and satisfies three axioms: it is associative, commutative, and idempotent. All three axioms are illustrated in the following proof.

Theorem 1. *Disjunction distributes through itself:*

$$(p \vee q) \vee r = (p \vee r) \vee (q \vee r)$$

Proof.

$$
\begin{aligned}
RHS &= p \vee (r \vee (q \vee r)) & \text{by associativity} \\
&= p \vee ((q \vee r) \vee r) & \text{by commutativity} \\
&= p \vee (q \vee (r \vee r)) & \text{by associativity} \\
&= p \vee (q \vee r) & \text{by idempotence} \\
&= LHS & \text{by associativity}
\end{aligned}
$$

Corollary 1. *Rightward distribution (follows by commutativity).*

Geometry. Geometry is recognised in Mathematics as an excellent way of gaining intuition about the meaning and the validity of algebraic axioms, proofs, conjectures, and theorems. The relevant geometric diagrams for Boolean algebra are familiar as Venn diagrams. For example, Fig. 1a illustrates the Venn diagram for disjunction.

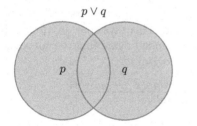
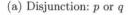
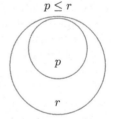

(a) Disjunction: p or q \qquad (b) Comparison: r is weaker than p

Fig. 1. Venn diagrams for disjunction and comparison.

Comparison (denoted as $\leq$). The most important comparison operator between terms of Boolean algebra is implication. It is written here as a simple less-than-or-equal sign ($\leq$). It is defined simply in terms of disjunction:

$$p \leq r \text{ is defined as } r = p \vee r$$

The comparison may be read in many ways: that p implies r, or that p is stronger than r, or that r is weaker than p. The definition is illustrated by a Venn diagram showing containment of the stronger left side p by the weaker right side r (see Fig. 1b).

Disjunction Is a Weakening Operator. An operator is defined as weakening if its result is always weaker than both of its operands. From Theorem 2 and Corollary 2 below, we conclude that the result of disjunction is always weaker than both of its operands. The proof of this again uses all three axioms.

Theorem 2. $p \leq p \vee r$

Proof.
$$p \vee r = (p \vee p) \vee r \quad \text{by idempotence}$$
$$= p \vee (p \vee r) \quad \text{by associativity}$$

The theorem follows by definition of $\leq$.

Corollary 2. $p \leq r \vee p$ *(by commutativity)*

Henceforth, we omit brackets around associative operators and proofs of theorems that follow by commutativity.

2.2 Deductive Logic

The axioms of algebra are restricted to single equations or comparisons between two algebraic terms. This makes algebraic reasoning quite simple, using only substitution of equals to deduce a new equation from two equations that have already been proved. The price of this simplicity is that proofs can get too long for comfort, and they can be quite difficult to find. To tackle these problems we need more powerful techniques, which are expressed as rules of logical deduction.

The Aristotelian Syllogism. A syllogism is a form of proof rule that has been taught for over two thousand years. It consists of two antecedents written above a line and one consequent written below the line. This says that any proof that contains both the antecedents can validly be extended by adding the consequent as its next line. A well-known example of a syllogism is:

$$\frac{\textit{All men are animals} \qquad \textit{All animals are mortal}}{\textit{All men are mortal}}$$

The use of syllogisms as a tool for reasoning can be dated back to the work of the ancient Greek philosopher Aristotle [32], who made a remarkable contribution to the history of human thought. He was the founder, director and a lecturer at a private academic institution in Athens. His lecture notes still survive. They deal with both the sciences and the humanities, and spanned almost the full range of human intellectual endeavour for the next two thousand years. The first application of syllogisms was probably in Biology, of which he is also recognised as the founding father. They are well adapted to deducing the consequences of his biological classifications.

A proof rule in algebra rather than biology is given in the following theorem.

Aristotle
(384–322 BC)

Theorem 3 (Proof by cases).

$$\frac{p \leq r \qquad q \leq r}{(p \vee q) \leq r}$$

Proof. Assuming the antecedents $r = p \vee r$ and $r = q \vee r$, we prove the consequent:

$$
\begin{aligned}
r &= r \vee r & &\text{by idempotence} \\
&= (p \vee r) \vee (q \vee r) & &\text{by substitution for each } r \\
&= (p \vee q) \vee r & &\text{by Theorem 1}
\end{aligned}
$$

The conclusion follows by definition of $\leq$.

In this proof, the assumption of the antecedents of the rule is justified by the general embargo which forbids use of the rule until the antecedents have already been proved.

A validated proof rule can also be used backwards to suggest a structure and strategy for a proof of a desired conjecture which matches the conclusion of the rule. Then the task of proof can be split into subtasks, one for each of the antecedents. Success of this strategy requires that each antecedent is in some way simpler than the conclusion. For example in the rule for proof by cases, the conclusion has a disjunction $p \vee q$ where the antecedents only contain a single operand, either p or q. The backward use is widely adopted in the search for proofs by computer.

Partial Orders. The well-known properties of an ordering in mathematics are usually defined by means of proof rules. The rules shown in the proof of Theorem 4 define the concept of a *partial order*. Each rule is proved by only one of the three axioms of disjunction. The first line shows how an axiom itself can be written as the consequent of a proof rule with no antecedents.

Theorem 4. *Comparison* $(\leq)$ *is a partial order.*

Proof. Comparison is:

reflexive: $$\frac{}{p \leq p}$$ (by idempotence)

transitive: $$\frac{p \leq q \quad q \leq r}{p \leq r}$$ (by associativity)

antisymmetric: $$\frac{p \leq q \quad q \leq p}{p = q}$$ (by commutativity)

Covariance (Monotonicity) of Disjunction. Covariance is the property of an operator that if either of its operands is strengthened, its result is also strengthened (or stays the same). Such an operator is said to respect the ordering of its operands. Covariance justifies the use of the comparison operator $\leq$ for substitution of one formula in another, just like the familiar rule of substitution of equals.

Theorem 5. *Disjunction is covariant (monotonic) with respect to $\leq$, that is:*

$$\frac{p \leq q}{p \vee r \leq q \vee r}$$

Proof. From the antecedent, transitivity of $\leq$, and weakening of disjunction, we have:

$$p \leq q \leq q \vee r \quad and \quad r \leq q \vee r$$

The consequent follows by the proof rule by cases.

Covariance is also a formal statement of a common principle of engineering reasoning. Suppose you replace a component in a product by one that has the same behaviour, but is claimed to be more reliable. The principle says that the product as a whole will be made more reliable by the replacement; or at least it will remain equally reliable. If the product is found in use to be less reliable than it was before the replacement, then the claimed extra reliability of the component is disproved.

2.3   Spatio-Temporal Logic

A theorem of Boolean algebra is used to state an universal truth, which remains true everywhere and forever. The ideas of temporal logic were explored by Aristotle and his successors, for reasoning about what may be true only during a certain interval of time (its duration), and in a certain area of space (its extent). A proposition describes all significant events occurring within its given duration and within its given extent. However, the logic does not allow any mention of a numeric measurement of the instant time or the point in space at which an event occurs. Thus a proposition in the logic can be true of many different regions of space and time.

Temporal logic was widely explored by philosophers and theologians in the middle ages. William of Occam (1287–1347), a Franciscan friar studying philosophy at Oxford, is considered to be one of the major figures of medieval thought. Unfortunately he got involved in church politics. He antagonised the pope in Rome, and was excommunicated from the Church in 1328. This was believed to condemn him to an eternity in hell. Fortunately, he was reprieved thirty years later. Occam's book on Logic, *Summa Logicae* (1323) included familiar operators of Boolean Algebra, augmented by operators that apply to propositions of spatial and temporal logic [25].

William of Occam
(1287-1347)

They include sequential composition *p then q*, written here with semicolon $(p; q)$, and *p while q*, written here with a single vertical bar $(p \mid q)$.

Geometric Diagrams. The propositions of Occam's spatio-temporal logic are best illustrated by two-dimensional geometric diagrams, with one axis representing time and the other representing space. As shown in Fig. 2a, the region described by a proposition p is represented by a rectangular box with the name p written in the top left corner. The box contains a finite set of discrete points, representing all the events that occurred in the region. The horizontal edges of the box represent the interval of time within which those events occur. The vertical edges represent the locations in space where the events occur. Figure 2b illustrates these two dimensions.

In Cartesian plane geometry, each point lies at the intersection of a vertical coordinate, shown here in gray, and a horizontal coordinate shown as a black arrow (Fig. 2b). Each point can therefore be identified by a pairing of a horizontal coordinate with a vertical coordinate. But the geometry shown here differs from this in that not all coordinate positions are occupied by a point. This is because in the description of the real world many or most coordinates are occupied by no event. Our diagrams are comparable to the output of a multiple pen recorder, for example the seismograms of geology and the cardiograms. Each horizontal line is the output of a single pen recording the value given by sensors in different locations. The events record significant changes in the value of the sensor.

In computer applications, the horizontal lines stand uniquely for a variable held in the memory of the computer. The events on a line represent assignments of potentially new values to the variables. The vertical lines are often drawn in later to explain a group of significant changes made simultaneously in many variables.

The sequential composition of p and q, denoted as $p;q$, starts with the start of p and ends with the end of q. Furthermore, q starts only when p ends. Figure 3a shows a diagram of the sequential composition of p and q. As before, the box is named by the term written in the top left corner. Every event in the composition is inside exactly one of the two operands. The vertical line between p and q is shared by both of them. It shows that time intervals of the two operands are immediately adjacent in time. The interval for the result is the set union of the interval for p and the interval for q .

The boxes with dotted edges at the corners of the $p; q$ contain no events. They are padding, needed to draw the result of composition as a box. To represent this padding we introduce an algebraic constant $\square$, read as '*skip*'.

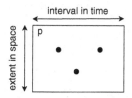

(a) Region describing p, with points representing events.

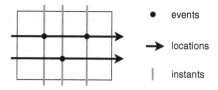

(b) Events occur within an interval in a given location in space.

Fig. 2. Propositions as two-dimensional geometric diagrams.

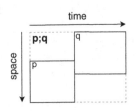

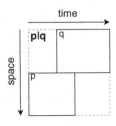

(a) Sequential composition.

(b) Concurrent composition.

Fig. 3. Spatio-temporal diagrams for sequential composition and concurrent composition.

The concurrent composition of p and q, denoted as $p \mid q$ and read as p *while* q, starts with the start of both p and q and ends with the end of both of them. Its duration is the maximum of their durations. Figure 3b shows a diagram of the concurrent composition of p and q. Its extent in space is the disjoint union of the extents of the operands. This means that no location can be shared by the concurrent components This embargo is the characteristic of O'Hearn's separation logic [23, 24, 30], which protects against the well-known problem of races in concurrent programs.

3   Sequential Composition

The algebraic axioms for sequential composition are:

- Sequential composition is associative and has the unit $\square$
- Sequential composition distributes through disjunction (both leftward and rightward):

$$p; (q \vee q') = p; q \vee p; q' \qquad \text{and} \qquad (q \vee q'); p = q; p \vee q'; p$$

Distribution justifies giving sequential composition a stronger precedence than disjunction. The associativity of sequential composition is evident from its diagram, and so is the unit law.

We now show how the algebraic axioms can be used to prove some rules.

Theorem 6 (Proof rule for sequential composition).

$$\frac{p; q \leq m \qquad m; r \leq t}{p; q; r \leq t}$$

Proof. Assuming the antecedents (1) $m = p; q \vee m$ and (2) $t = \boxed{m}; r \vee t$, we prove the consequent:

$$
\begin{aligned}
t &= \boxed{(p; q \vee m)}; r \vee t & &\text{substitute (1) in (2)} \\
t &= p; q; \boxed{r} \vee (m; \boxed{r} \vee t) & &\text{; distributes through } \vee \\
t &= p; q; r \vee t & &\text{substitute back by (2)}
\end{aligned}
$$

This proof rule is used for decomposing its consequent into two parts, each of which has only three operands instead of four. Each antecedent is in this way simpler than the consequent, whose proof can therefore be constructed by divide and conquer.

Rules of Consequence. The following corollaries are a consequence of Theorem 6.

Corollary 3.

$$\frac{p \leq m \qquad m; r \leq t}{p; r \leq t}$$

Corollary 4.

$$\frac{p; q \leq m \qquad m \leq t}{p; q \leq t}$$

Proof. Corollary 3: by substitution of q by $\square$. Corollary 4: by substitution of r by $\square$.

3.1 Hoare Triples

Consider the proposition $p; q \leq r$. It means that if p describes the interval from the start of r to the start of q, and q describes the interval from the end of p to the end of r, then r correctly describes the whole of $p; q$. This is the intended meaning of the Hoare triple [14]. Therefore, we define:

$$\{p\}\, q\, \{r\} \stackrel{\text{def}}{=} p; q \leq r$$

This definition allows p and r to be arbitrary programs—a generalisation of the original formulation of Hoare logic, in which p and r are required to be assertions.

3.2 Verification Rules for Sequential Composition

By substitution of the definition of triple into the proof rule for sequential composition (Theorem 6), we obtain the Hoare rule for sequential composition:

$$\frac{\{p\}\, q\, \{m\} \qquad \{m\}\, r\, \{t\}}{\{p\}\, q\,;\, r\, \{t\}}$$

From Corollaries 3 and 4, we obtain the Hoare Rules of Consequence:

$$\frac{p \leq m \qquad \{m\}\, r\, \{t\}}{\{p\}\, r\, \{t\}} \qquad\qquad \frac{\{p\}\, q\, \{m\} \qquad m \leq t}{\{p\}\, q\, \{t\}}$$

3.3   Milner Transition

Robin Milner defined CCS [19], a theory of programming which is now widely used in specifying how an implementation should generate a single execution of a given program r. The Milner transition defined here, and denoted $r \xrightarrow{p} q$, states that r can be executed by executing p first, saving q as a continuation for subsequent execution. (Other executions may begin with an initial step different from p). But this is exactly the meaning of the same comparison that we used to define the Hoare triple. We thus define:

$$r \xrightarrow{p} q \overset{\text{def}}{=} p; q \leq r$$

Thus the two calculi are identical, and all theorems of one can be translated letter by letter from the corresponding theorem of the other. For example, in Milner's notation the rule for sequential composition and its corollaries are

$$\frac{r \xrightarrow{p} m \qquad m \xrightarrow{q} t}{r \xrightarrow{p;q} t}$$

$$\frac{m \leq r \qquad m \xrightarrow{q} t}{r \xrightarrow{q} t} \qquad \frac{r \xrightarrow{p} m \qquad t \leq m}{r \xrightarrow{p} t}$$

These corollaries play the role of the structural equivalence, which Milner introduced into the definition of concurrent programming languages (with $\equiv$ replaced by $\leq$) [20].

4   Concurrent Composition

Concurrent composition has the same laws as sequential composition. An additional *interchange axiom* permits a concurrent program to be executed sequentially by interleaving. The algebraic axioms are:

– Concurrent composition is associative and has unit $\square$
– Concurrent composition distributes through disjunction
– Interchange axiom: $(p \mid q); (p' \mid q') \leq (p; p') \mid (q; q')$

We omit the commonly cited commutativity law for concurrency since it can be introduced later, whenever needed. The interchange law gets its name because it interchanges operators and variables when passing from one side of the comparison to the other. Note how the RHS and LHS differ by interchange of operators (; interchanged with $\mid$) and of operands (p' with q).

4.1   Interchange

The two following elementary corollaries of interchange show that a concurrent composition can be strengthened by sequential execution of its operands in either order:

$$p; q' \leq p \mid q' \qquad \text{by interchange with } p' = q = \square$$
$$q; p' \leq p' \mid q \qquad \text{similarly, with } q' = p = \square$$

From these two properties and the proof rule by cases, we obtain:

$$p; q \vee q; p \ \leq \ p \mid q$$

This means that concurrent composition is weaker than the disjunction of these alternative orderings. We will now show by example that the interchange law generalises this interleaving to operands containing any number of operators.

We start with what are known as small interchange laws.

Theorem 7 (Small interchange laws).

$$
\begin{aligned}
p; (p' \mid q') &\leq (p; p') \mid q' & q &= \square \\
q; (p' \mid q') &\leq p' \mid (q; q') & p &= \square \\
(p \mid q); q' &\leq p \mid (q; q') & p' &= \square \\
(p \mid q); p' &\leq (p; p') \mid q & q' &= \square
\end{aligned}
$$

Proof. All four are proved from the interchange axiom, by substitution of $\square$ for a different variable.

The above six corollaries are called frame laws in separation logic. They adapt the interchange law to cases with just two or three operands. Successive application of the frame laws can strengthen any term with two or three operands to a form not containing any concurrency. The following is an example derivation:

$$p; q; q' \leq (p \mid q); q' \leq p \mid (q; q')$$

4.2   Basic Principle of Concurrent Programming

We now show how to interleave longer strings. Let x, y, z, w, a, b, c, d be characters representing single events. Let us omit ";" in strings except for emphasis. Thus:

$$xyzw = x; y; z; w$$

The interchange law itself extends this principle to arbitrary terms, with many concurrent compositions, as the following example shows:

$abcd \mid xyzw$	is the RHS of interchange
$\geq (\, a; bcd\,) \mid (xy; zw)$	associativity (twice)
$\geq (\, a \mid xy); (\, bcd \mid zw)$	interchange
$\geq (\, a \mid x; y); (\, b; cd \mid zw)$	associativity (twice)
$\geq (\, a \mid x); y; (\, b \mid zw); cd$	frame laws (twice)
$\geq x\, a\, yz\, b\, w\, cd$	similarly

In the first line of this derivation, the characters of the left operand of concurrency have been highlighted; and the same characters are highlighted in subsequent lines. This conveys the important intuition that the order of characters in each sequential substring is preserved throughout. The same applies to the original right operand. Furthermore, each line splits some of the substrings of the previous line into two substrings. When all the highlighted substrings are of length 1, the first corollary can eliminate the concurrency. This shows that any chain of calculation using the interchange law must terminate.

A basic principle of concurrent programming states that every concurrent program can be simulated by a sequential program. Without this principle, it would have been impossible to exploit concurrency in general-purpose libraries and class declarations. The principle was proved for Turing machines by the design of a normal sequential Turing machine that could interpret any program run by multiple machines [27]. Our result is that any concurrent program can be translated by algebraic transformations for execution by a purely sequential machine. A direct algebraic proof is much simpler than a proof by interpretation. The result is also more useful because it can be applied to arbitrary sub-terms of a term. Thus the explosive increase in length of most reductions to normal form can generally be avoided.

4.3   Unifying Theories of Concurrency

The basic concurrency rule of separation logic was formulated by Peter O'Hearn in Hoare Logic. When translated to our algebraic notation it gives the following proof rule.

Interchange Rule (O'Hearn).

$$\frac{p;q \leq r \qquad p';q' \leq r'}{(p \mid p');(q \mid q') \leq (r \mid r')}$$

His frame rule similarly translates to one of the frame laws of Theorem 7.

Just as the sequential rule is derived from the sequential axioms in Sect. 3, the Interchange Rule is derivable from the Interchange Axiom.

Theorem 8. *The Interchange Axiom implies the Interchange Rule.*

Proof. Assume the antecedents of the interchange rule:

$$p;q \leq r \qquad \text{and} \qquad p';q' \leq r'$$

$$(p \mid p'); (q \mid q') \leq \boxed{(p;q) \mid (p';q')} \qquad \text{Covariance of } \mid \text{ twice:}$$

$$\boxed{(p;q) \mid (p';q')} \leq (r \mid r')$$
and transitivity of $\leq$

$$(p \mid p'); (q \mid q') \leq (r \mid r')$$

Conclusion: $\qquad \dfrac{p;q \leq \boxed{r} \qquad p';q' \leq \boxed{r'}}{(p \mid p'); (q \mid q') \leq \boxed{(r \mid r')}} \qquad$ the interchange rule

Surprisingly, the implication also holds in the reverse direction.

Theorem 9. *The Interchange Rule implies the Interchange Axiom.*

Proof. We start by assuming the interchange rule. Since it is a general rule, we can replace consistently all occurrences of each of its variables by anything we like.

$$\dfrac{p;q \leq \boxed{r} \qquad p';q' \leq \boxed{r'}}{(p \mid p'); (q \mid q') \leq \boxed{(r \mid r')}} \qquad \text{replace } \boxed{r} \text{ by } \boxed{p;q}$$
and $\boxed{r'}$ by $\boxed{p';q'}$

$$\dfrac{p;q \leq \boxed{p;q} \qquad p';q' \leq \boxed{p';q'}}{(p \mid p'); (q \mid q') \leq (p;q \mid p';q')} \qquad \begin{array}{l}\text{both antecedents are true}\\ \text{by reflexivity of } \leq\end{array}$$

Conclusion: $(p \mid p'); (q \mid q') \leq (p;q) \mid (p';q') \qquad$ the interchange axiom

Summary. We have extended to concurrency the unification between Hoare Triples and Milner Transitions that was achieved for sequentiality in Sect. 3.

Theorem 10. *The following three rules are logically equivalent.*

$$\dfrac{p;q \leq r \qquad p';q' \leq r'}{(p \mid p'); (q \mid q') \leq (r \mid r')} \qquad \textit{The Interchange Rule}$$

$$\dfrac{\{p\}\, q\, \{r\} \qquad \{p'\}\, q'\, \{r'\}}{\{(p \mid p')\}\, q \mid q'\, \{(r \mid r')\}} \qquad \textit{Translated to Hoare Triples}$$

$$\dfrac{r \xrightarrow{p} q \qquad r' \xrightarrow{p'} q'}{(r \mid r') \xrightarrow{(p \mid p')} (q \mid q')} \qquad \textit{Translated to Milner transitions}$$

The third rule is just the rule for concurrency in Milner's CCS, as formulated in the so-called 'big-step' version of operational semantics. It is interpreted as stating:

> To execute a concurrent composition of two sequential operands, split each operand into two sequential parts. Then start by executing the first part of both operands concurrently, and conclude by executing the second parts.

The unification of two widely accepted theories of programming is presented as strong evidence that our algebraic axioms are actually applicable to familiar programming languages implemented on computers of the present day. Many interpreters and compilers for programming languages are specified by an operational semantics expressed as Milner Transitions. Most program analysers and proof tools for sequential languages follow a verification semantics expressed as Hoare Triples. Many papers in the Theory of Programming prove the consistency between these two 'rival' theories for particular languages. Algebra unifies the theories, by proofs which could be understood or even discovered (under guidance) by CS students in their practical programming courses.

5 Related Work

This section surveys evidence for the validity of the three theses listed in the Introduction.

1. The biographies in this paper of Aristotle, Boole, and Occam are only a small selection of those who have contributed to the basic ideas of Computer Science, long before computers were available to put them into practice. Further examples are Euclid and Descartes for Geometry, Al-Khawarismi and Leibniz for Algebra, and Russel and Gödel for Logic. Their biographies may be found in Wikipedia. More recent pioneers are treated in [8].
2. Considerable experience has been accumulated of the effectiveness of teaching the Theory of Programming as part of practical degree courses in Computer Science. For example, in [29], the authors show how teaching concurrency and verification together can reinforce each other and enable deeper understanding and application. They suggest that concurrency should be taught as early as possible and they introduce a new workflow methodology that is based on existing concurrency models (CSP, π-calculus), on the model checker FDR that generates counter-example traces that show causes of errors, and on programming languages/libraries (occam-π, Go, JCSP, ProcessJ) that allow executable systems within these models.

   Another interesting example is the experimental course in "(In-)Formal Methods" [21], where invariants, assertions, and static reasoning are introduced. The author argues that the ideal place for an informal-methods course is the second half of first year, because at that point students already understand that "programming is easy, but programming correctly is very hard".

Further proposals to introduce invariants and assertions as part of the introductory Computer Science curriculum, even at pre-university level, are presented in [10] and [11]. In [10], a programme focused on algorithmic problem
solving and calculational reasoning is proposed. In [11], an experiment is presented where students specify algorithmic problems in Alloy [17] and reason
about problems in an algebraic and calculational way. It has been argued that
students seem to prefer and understand better calculational proofs [9]. Calculational proofs are commonly used in the functional programming community
to demonstrate algorithm correctness [4, 16]. Recent tool support shows that
this style can have impact in practical functional programming [33]. An application of relational calculus to software verification is presented in [26],
illustrated with a case study on developing a reliable FLASH filesystem for
in-flight software. It combines the pragmatism of Alloy [17] with the Algebra
of Programming presented in [5].

3. The introduction of formal methods in practical programming has accelerated in recent years. Regarding practical verification, there have been several attempts at building languages and systems that support verification,
providing the ability to specify preconditions, postconditions, assertions, and
invariants. ESC/Java [12] and Spec# [3] build on existing languages, Java and
C#, respectively. Dafny [18] is a programming language with built-in specification constructs. The Dafny static program verifier can be used to verify
the functional correctness of programs. Dafny has been extensively used in
teaching. Whiley [28] is a programming language designed from scratch in
conjunction with a verifying compiler. SOCOS [2] is a programming environment that applies Invariant Based Programming [1], a visual and practical
program construction and verification methodology. The Java+ITP [31] was
used as a teaching tool at the University of Illinois at Urbana-Champaign to
teach graduate students and seniors the essential ideas of algebraic semantics
and Hoare logic. A recent case of success in industry is Infer[1] [7], a static
analyzer based on separation logic [30] adopted and being developed by Facebook. Infer has been used in a 4th-year MEng and MSc course on separation
logic at the Department of Computing, Imperial College London[2].

6   Conclusion

We hope that this article has contributed to the challenge posed by Carroll
Morgan that we mentioned in the Introduction. We also hope to have made
the case that current achievements in teaching sequential programming can be
extended to concurrent programming.

---

[1] Infer static analyzer website: https://fbinfer.com.
[2] Course link: https://vtss.doc.ic.ac.uk/teaching/InferLab.html.

The theory has been further extended to object oriented programming in [15]. These extensions will require new textbooks and extension and combination of existing tools. The creation of an environment that effectively combines the experience and tools already available is an open challenge. Ideally, the environment should allow students to work at different levels of abstraction and should unify interfaces and techniques from existing tools, such as Alloy Analyzer [17] and Isabelle/UTP [13]. Since this environment is to be used in a teaching environment, we do not have the problem of scale; however, feedback must be given quickly to students (and preferably in a graphical form). The approach described in [29] is an excellent example of how a model-checker for concurrency can be integrated with a testing tool. We believe it would be fruitful if tool-builders and users adopted a similar approach, integrating their tools and ideas into this system and other rival verification platforms. Tools such as the theorem prover Lean [22] seem to provide a promising basis for further developments.

References

1. Back, R.-J.: Invariant based programming: basic approach and teaching experiences. Formal Asp. Comput. **21**(3), 227–244 (2009)
2. Back, R.-J., Eriksson, J., Mannila, L.: Teaching the construction of correct programs using invariant based programming. In: Proceedings of the 3rd South-East European Workshop on Formal Methods (2007)
3. Barnett, M., Leino, K.R.M., Schulte, W.: The Spec# programming system: an overview. In: Barthe, G., Burdy, L., Huisman, M., Lanet, J.-L., Muntean, T. (eds.) CASSIS 2004. LNCS, vol. 3362, pp. 49–69. Springer, Heidelberg (2005). https://doi.org/10.1007/978-3-540-30569-9_3
4. Bird, R.: Pearls of Functional Algorithm Design. Cambridge University Press, Cambridge (2010)
5. Bird, R., De Moor, O.: The algebra of programming. In: NATO ASI DPD, pp. 167–203 (1996)
6. Boole, G.: An Investigation of the Laws of Thought: On Which Are Founded the Mathematical Theories of Logic and Probabilities. Dover Publications, New York (1854)
7. Calcagno, C., Distefano, D.: Infer: an automatic program verifier for memory safety of C programs. In: Bobaru, M., Havelund, K., Holzmann, G.J., Joshi, R. (eds.) NFM 2011. LNCS, vol. 6617, pp. 459–465. Springer, Heidelberg (2011). https://doi.org/10.1007/978-3-642-20398-5_33
8. Davis, M.: Engines of Logic: Mathematicians and the Origin of the Computer. WW Norton & Co., Inc., New York (2001)
9. Ferreira, J.F., Mendes, A.: Students' feedback on teaching mathematics through the calculational method. In: 2009 39th IEEE Frontiers in Education Conference, pp. 1–6. IEEE (2009)
10. Ferreira, J.F., Mendes, A., Backhouse, R., Barbosa, L.S.: Which mathematics for the information society? In: Gibbons, J., Oliveira, J.N. (eds.) TFM 2009. LNCS, vol. 5846, pp. 39–56. Springer, Heidelberg (2009). https://doi.org/10.1007/978-3-642-04912-5_4

11. Ferreira, J.F., et al.: Logic training through algorithmic problem solving. In: Blackburn, P., van Ditmarsch, H., Manzano, M., Soler-Toscano, F. (eds.) TICTTL 2011. LNCS (LNAI), vol. 6680, pp. 62–69. Springer, Heidelberg (2011). https://doi.org/10.1007/978-3-642-21350-2_8

12. Flanagan, C., et al.: Extended static checking for Java. ACM SIGPLAN Not. 37(5), 234–245 (2002)

13. Foster, S., Zeyda, F., Woodcock, J.: Isabelle/UTP: a mechanised theory engineering framework. In: Naumann, D. (ed.) UTP 2014. LNCS, vol. 8963, pp. 21–41. Springer, Cham (2015). https://doi.org/10.1007/978-3-319-14806-9_2

14. Antony, C., Hoare, C.A.R.: An axiomatic basis for computer programming. Commun. ACM 12(10), 576–580 (1969)

15. Hoare, T., Struth, G., Woodcock, J.: A calculus of space, time, and causality: its algebra, geometry, logic. In: Ribeiro, P., Sampaio, A. (eds.) UTP 2019. LNCS, vol. 11885, pp. 3–21. Springer, Cham (2019)

16. Hutton, G.: Programming in Haskell. Cambridge University Press, Cambridge (2016)

17. Jackson, D.: Alloy: a lightweight object modelling notation. ACM Trans. Softw. Eng. Methodol. (TOSEM) 11(2), 256–290 (2002)

18. Leino, K.R.M.: Dafny: an automatic program verifier for functional correctness. In: Clarke, E.M., Voronkov, A. (eds.) LPAR 2010. LNCS (LNAI), vol. 6355, pp. 348–370. Springer, Heidelberg (2010). https://doi.org/10.1007/978-3-642-17511-4_20

19. Milner, R. (ed.): A Calculus of Communicating Systems. LNCS, vol. 92. Springer, Heidelberg (1980). https://doi.org/10.1007/3-540-10235-3

20. Milner, R.: Communicating and Mobile Systems: The Pi Calculus. Cambridge University Press, Cambridge (1999)

21. Morgan, C.: (In-)formal methods: the lost art. In: Liu, Z., Zhang, Z. (eds.) SETSS 2014. LNCS, vol. 9506, pp. 1–79. Springer, Cham (2016). https://doi.org/10.1007/978-3-319-29628-9_1

22. de Moura, L., Kong, S., Avigad, J., van Doorn, F., von Raumer, J.: The Lean theorem prover (system description). In: Felty, A.P., Middeldorp, A. (eds.) CADE 2015. LNCS (LNAI), vol. 9195, pp. 378–388. Springer, Cham (2015). https://doi.org/10.1007/978-3-319-21401-6_26

23. O'Hearn, P.: Resources, concurrency, and local reasoning. Theor. Comput. Sci. 375(1–3), 271–307 (2007)

24. O'Hearn, P., Reynolds, J., Yang, H.: Local reasoning about programs that alter data structures. In: Fribourg, L. (ed.) CSL 2001. LNCS, vol. 2142, pp. 1–19. Springer, Heidelberg (2001). https://doi.org/10.1007/3-540-44802-0_1

25. William of Ockham: Ockham's Theory of Propositions: Part II of the Summa Logicae. University of Notre Dame Press (1980). Translated by Alfred J. Freddoso and Henry Schuurman

26. Oliveira, J.N., Ferreira, M.A.: Alloy meets the algebra of programming: a case study. IEEE Trans. Softw. Eng. 39(3), 305–326 (2012)

27. Papadimitriou, C.H.: Computational Complexity. Wiley, Hoboken (2003)

28. Pearce, D.J., Groves, L.: Whiley: a platform for research in software verification. In: Erwig, M., Paige, R.F., Van Wyk, E. (eds.) SLE 2013. LNCS, vol. 8225, pp. 238–248. Springer, Cham (2013). https://doi.org/10.1007/978-3-319-02654-1_13

29. Pedersen, J.B., Welch, P.H.: The symbiosis of concurrency and verification: teaching and case studies. Formal Asp. Comput. 30(2), 239–277 (2018)

30. Reynolds, J.C.: Separation logic: a logic for shared mutable data structures. In: Proceedings 17th Annual IEEE Symposium on Logic in Computer Science, pp. 55–74. IEEE (2002)
31. Sasse, R., Meseguer, J.: Java+ ITP: a verification tool based on Hoare logic and algebraic semantics. Electron. Notes Theor. Comput. Sci. **176**(4), 29–46 (2007)
32. Smith, R.: Prior Analytics. Hackett Publishing, Indianapolis (1989)
33. Vazou, N., Breitner, J., Kunkel, R., Van Horn, D., Hutton, G.: Theorem proving for all: equational reasoning in liquid Haskell (functional pearl). In: Proceedings of the 11th ACM SIGPLAN International Symposium on Haskell, pp. 132–144. ACM (2018)

Teaching Program Verification

Teaching Deductive Verification Through Frama-C and SPARK for Non Computer Scientists

Léo Creuse[1], Claire Dross[2], Christophe Garion[1(✉)] , Jérôme Hugues[1] ,
and Joffrey Huguet[1]

[1] ISAE-SUPAERO, Université de Toulouse, Toulouse, France
{creuse,huguet}@student.isae-supaero.fr, {garion,hugues}@isae-supaero.fr
[2] AdaCore, Paris, France
dross@adacore.com

Abstract. Deductive verification of software is a formal method that is usually taught in Computer Science curriculum. But how can students with no strong background in Computer Science be exposed to such a technique? We present in this paper two experiments made at ISAE-SUPAERO, an engineering program focusing on aerospace industry. The first one is a classic lecture introducing deductive methods through the Frama-C platform or the SPARK programming language. The second one is the production by two undergraduate students of a complete guide on how to prove complex algorithms with SPARK. Both experiments showed that students with no previous knowledge of formal methods nor theoretical Computer Science may learn deductive methods efficiently with bottom-up approaches in which they are quickly confronted to tools and practical sessions.

1 Introduction

Formal methods are usually taught in Computer Science curriculum where students have a good background in theoretical Computer Science. But how to teach formal methods in a more "hostile" environment, for instance where students only have a minimal background in Computer Science? ISAE-SUPAERO engineering program is such an environment, as students are exposed to a small amount of hours in Computer Science and almost nothing in theoretical Computer Science.

However, as ISAE-SUPAERO is mainly oriented to aerospace industry, we think that introducing formal methods is crucial, at least for students choosing the Critical Systems final year major. In order to do so, we have first introduced a classic course in this major with two tracks on deductive verification: one using Frama-C and its plugin WP to verify C programs and another one using the SPARK programming language and its associated tool GNATProve. These two tracks are taught with different educational methods. The first one uses a classic top-down approach with theory on deductive verification presented before using

B. Dongol et al. (Eds.): FMTea 2019, LNCS 11758, pp. 23–36, 2019.
https://doi.org/10.1007/978-3-030-32441-4_2

tools, whereas the second one uses a bottom-up approach in which students uses SPARK from the beginning of the course and prove more and more complex properties on their programs.

We have also asked two undergraduate students to develop a complete guide on how to use SPARK to prove classic Computer Science algorithms (algorithms on arrays, binary heaps, sorting algorithms). This is inspired by *ACSL by Example*, a similar guide developed for the ACSL specification language for C programs, hence the name of our guide, *SPARK by Example*. The development of *SPARK by Example* showed us some pitfalls and difficulties encountered by beginners and we hope that the resulting document may help to understand how to specify and prove SPARK programs.

This paper is organized as follows. Section 2 presents ISAE-SUPAERO engineering program and why it is difficult to teach formal methods in this context. Section 3 presents a last year course on formal methods and focuses on tracks on deductive verification through Frama-C and SPARK. Section 4 presents *SPARK by Example*, a complete guide on how to prove 50 algorithms taken from the C++ STL. Finally, Sect. 5 is dedicated to conclusion and perspectives.

2   Why Is It Difficult to Teach Formal Methods at ISAE-SUPAERO?

2.1   The ISAE-SUPAERO Engineering Program

ISAE-SUPAERO is one of the leading French "Grandes Écoles" and is mostly dedicated to the aerospace industrial sector even if half of its students begin their career in other domains (research, energy, bank, IT, ...). Before entering ISAE-SUPAERO, students are selected through a national competitive exam for which they attend preparatory classes during two years. The ISAE-SUPAERO engineering program [14] is a three-year program during which students acquire common scientific and non-scientific background on aerospace industry and also choose elective courses to prepare themselves for their career. The three years are split into 6 semesters whose content is presented on table 1.

Table 1. Content of semesters in ISAE-SUPAERO engineering program

Semester	Content
S1	Common core
S2	Elective courses & projects
S3	Common core
S4	Elective courses & projects
S5	Field of application (140 h) & Major of expertise (240 h)

Figure 1 show the ratio of Computer Science oriented courses in the S1 and S3 common core courses. There are only three Computer Science courses:

- A 40 h course on Algorithm and Programming in S1. This course focuses on basic algorithms, classic data structures (linked lists, binary search trees, graphs) and C is the associated programming language. No formal methods are discussed during the lecture, but algebraic specifications are used to specify data structures.
- A 40 h course on Object-Oriented Design and Programming in S3, focusing on object-oriented design principles and programming in Java.
- A 10 h course on Integer Linear Programming in S3 during which complexity theory is tackled up to NP-completeness.

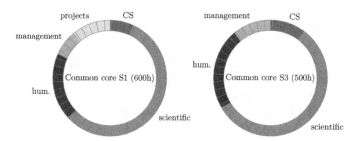

Fig. 1. Computer science part in S1 and S3 semesters

S2 and S4 are mainly dedicated to 30-h elective courses and some of them are Computer Science oriented, e.g. courses on functional and logic programming languages, implementation of control software in C or avionics architecture. However, none of the proposed course provides the students theoretical foundations of Computer Science nor formal methods.

The Computer Science, Telecommunications and Networks major of expertise in S5 have a Computer Science 240-h track mainly oriented towards Critical Systems Architecture. This tracks is composed of 6 courses presented on Table 2. The Model-Based Engineering course is composed of two parts: a 38-h part on SysML and SCADE and a 17-h part on formal methods.

2.2   The Challenges

In order to expose the students of the Critical Systems Architecture major to formal methods, we face several challenges:

- First, we cannot rely on elective courses proposed during S2 and S4 as the program does not enforce specific elective courses for each S5 major. We can only suppose that the students have been exposed to the 90 h of Computer Science courses in the common core, mainly on programming.
- Even if the scientific common core in S1 and S3 is rather large, non Computer Science scientific courses mainly rely on continuous mathematics, which are

Table 2. The courses of the critical systems architecture major

Course	Title	Volume
FITR301	Network and computer architecture	40 h
FITR302	Security	24 h
FITR303	Real-time systems	65 h
FITR304	Model-based engineering	55 h
FITR305	Distributed systems	35 h
FITR306	Conferences	21 h

not necessarily useful for formal methods. Particularly, students lack background on mathematical logic, calculability theory, programming languages semantics etc. On the other hand, students having attended the preparatory courses have a strong background in "classic" mathematics and are comfortable with mathematical proofs.
– There is only a 17-h slot in S5 to expose students to formal methods.

2.3 Why Teaching Formal Methods at ISAE-SUPAERO?

Given these pitfalls, one can wonder why teaching formal methods at ISAE-SUPAERO, particularly with a 17-h slot? We identify two points that justify this decision:

– As the main industrial sector of SUPAERO is aerospace, an introduction to formal methods seems legitimate, particularly for the students attending the Critical Systems Architecture major.
– Formal methods give more visibility to Computer Science as a *science*. As most students are exposed to Computer Science through programming courses, a non negligible part of them sees Computer Science as a technology, not a science. Notice that we do not consider programming as an minor part of a Computer Science curriculum, but it should not be the only one.

3   Introducing Formal Methods in 17 h

Instead of designing a course in which students are exposed to several formal methods, we decided to give students a brief introduction on the subject and then let them choose to learn a particular formal method through a dedicated track. As there are about 16 students in the major, this means that each track will be followed by 4 students. The tracks are the following:

– A track on *model-checking* through LTL and CTL, in which students model a system and prove some temporal properties on it

- A track on *abstract interpretation,* in which students implement an abstract interpreter on a tiny imperative language
- A track on *deductive methods* using SPARK
- A track on *deductive methods* using Frama-C.

The 17 h are divided as follows:

1. A 2-h introduction lecture during which we present what are formal methods, and what are their industrial uses. A small introduction to programming languages semantics is also done through a toy imperative language via denotational, operational and axiomatic semantics.
2. Each track has then six 2-h sessions to work on the corresponding technique. These sessions mix theoretical concepts and practical exercices.
3. Each track evaluates its students through a group project.
4. Each students group has then 30 min to present to the other groups the principles of the method they used, their results, the difficulties they encounter etc.
5. A 2-h industrial feedback is then done on how (aerospace) industry uses formal methods.

In the following, we will focus on the two tracks on deductive methods, particularly because they use two different pedagogical approaches.

3.1 Deductive Verification with Frama-C and SPARK

Deductive verification of programs is a formal method or technique that translates the problem of verifying a program annotated by assertions, invariants and contracts to the satisfiability problem of a particular mathematical logic formulas. Deductive verification relies on early work by Hoare [13], Floyd [12] and Dijkstra [9]. We are using two platforms for deductive verification:

- The Frama-C platform [16] with its WP plugin for deductive verification and the ACSL specification language [6]. Frama-C is dedicated to the analysis of C programs.
- The SPARK language [4] with its associated tool GNATProve. SPARK is a subset of the Ada programming language targeting formal verification.

3.2 A Top-Down Approach with Frama-C

The Frama-C track is built on a classic top-down approach: the students are first exposed to theory and then use the Frama-C tool. The first 2 h are dedicated to proof theory, particularly on formal systems for propositional and first-order logics. Floyd-Hoare logic is then presented and an introduction to weakest precondition calculus is done in 2 h. Students have then to manually annotate small imperative programs (e.g. factorial, euclidian division, greatest common divisor) to understand how weakest precondition works. Let us take for instance a factorial algorithm represented on Listing 1.1. The students have to discover the loop invariant and variants for the while loop and then use Floyd-Hoare logic rules to annotate each line of the program with assertions about memory. These small exercises are rather useful as they show the students that most of the assertions can be computed automatically, that proof obligations need to reason with particular theories in order to be discharged and that invariants are crucial to prove the expected postconditions but are sometimes difficult to find.

Listing 1.1. A simple imperative program computing factorial

```
{N ≥ 0}
K := 0
F := 1
while (K ≠ N) do
      K := K + 1;
      F := F * K
od
{F = N!}
```

The next 3 sessions are dedicated to a presentation of the Frama-C platform and its WP plugin for deductive verification. The exposure to Frama-C and ACSL is done gradually through small C programs that students have to prove or for which specifications are incomplete or wrong and that students have to debug. For instance, Listing 1.2 presents the previous factorial algorithm written in C with an axiomatic for factorial. Pointers are also tackled, particularly memory separation, which is more difficult.

Listing 1.2. A C program computing factorial

```
/*@ axiomatic factorial {
  @     predicate is_fact(integer n, integer r);
  @     axiom zero:
  @        is_fact(0, 1);
  @     axiom succ:
  @        \forall integer n, integer f;
  @        n>0 ==> is_fact(n-1, f) ==> is_fact(n, f*n);
  @
  @     logic integer fact(integer i);
  @     axiom fact1: \forall integer n;
  @                  is_fact(n, fact(n));
  @     axiom fact2: \forall integer n, integer f;
  @                  is_fact(n, f) ==> f == fact(n);
  @ }
  @*/

int factorial(int n) {
      int k = 1;
      int f = 1;
```

```
while (k <= n) {
    f = f * k;
    k = k + 1;
}

return f;
}
```

The project the students have to work on is rather classic: it consists in specifying, implementing and proving a small library on strings consisting of three functions:

```
int strlen(const char *str);
void strsubstring(char *dst, const char *src,
                  int start, int length);
void strappend(char *dst, const char *src);
```

3.3   A Bottom-Up Approach with SPARK

The track on deductive methods with SPARK is guided by a different methodology proposed by [5]. This approach is based on different levels of verification applied to SPARK code. The students are gradually exposed to the SPARK language and associated tools:

- On stone level, students have to write valid SPARK code. As a subset of the Ada language, SPARK has some limitations compared to Ada. As students have not been exposed to Ada programming before this course, attaining stone level is just learning a new language syntax and rational for them.
- In order to validate bronze level, SPARK programs must be such that there is no uninitialized variables nor interferences between parameters nor global variables.
- Silver level corresponds to Absence of RunTime Errors (AoRTE). These errors are typically overflows or underflows on integers, or accesses outside the range of an array. In order to validate SPARK programs on the silver level, students have to add preconditions to their functions, e.g. to restrict the possible value of a parameter to avoid possible overflows. They also have to write postconditions when several functions are interacting and are thus initiated to modular proof.
  A typical exercise is to specify a simple stack implementing with an array and offering Initialize , Push and Pop functions. Of course, manipulating an array may lead to illegal accesses, and the students have to add preconditions and postconditions to avoid them.
- Finally, gold level corresponds to functional correctness of SPARK programs. In order to prove that a program is correct, students have to understand how to write more complex contracts for functions and to add invariants and variants if some functions include loops through simple exercises. For instance, Listing 1.3 presents a Find_Int_Sqrt function implementing integer square root algorithm for which the students had to find a loop invariant.

Listing 1.3. A SPARK program computing integer square root

```
function Find_Int_Sqrt (N : in Natural) return Natural
is
    Lower, Upper, Middle: Natural;
    Maximum_Root: constant Natural := 46341;
begin
    Lower := 0;

    if N >= Maximum_Root then
        Upper := Maximum_Root;
    else
        Upper := N + 1;
    end if;

    loop
        --  Add a pragma Loop_Invariant and a
        --  pragma Loop_Variant here.

        exit when Lower + 1 = Upper;
        Middle := (Lower + Upper) / 2;
        if Middle * Middle > N then
            Upper := Middle;
        else
            Lower := Middle;
        end if;
    end loop;
    return Lower;
end Find_Int_Sqrt;
```

The project proposed to the students was to prove a small part of the Ada.Strings.Fixed GNAT standard library. This library consists in functions and procedures manipulating fixed size strings. For instance, the Index function shown on listing 1.4 searches for the first or last occurrence of any of a set characters, or any of the complement of a set of characters.

Listing 1.4. The index function from the GNAT standard library

```
function Index
    (Source : String;
     Set    : Maps.Character_Set;
     Test   : Membership := Inside;
     Going  : Direction  := Forward) return Natural;
```

Students were given 12 functions with their complete implementation and some specification written in natural language. They had to formally specify and prove these 12 functions.

3.4   Comparison of Both Approaches

First, let us notice that both groups of students manage to complete the assigned projects. Even if there were more functions to specify and prove in the SPARK project, the C functions used pointers as parameters, which was difficult to handle for students (particularly the problem of memory separation and the amount of specification needed to prove programs using pointers).

Of course, a top-down approach in such a small amount of time is not efficient. Starting from theory, particularly proof theory, takes a lot of time and there is not enough time to manipulate Frama-C. On the contrary, the bottom-up approach chosen for the SPARK track was more efficient: hands-on session from the beginning of the track is clearly an advantage as students immediately see the benefits of deductive verification on one particular aspect (control flow, AoRTE, functional correctness). Student presentations also shown that students having attended the SPARK track had more hindsight than the ones having attended the Frama-C track.

Finally, notice that the comparison between both approaches may be biased as C is a more difficult language than SPARK, particularly when using pointers.

4   The SPARK by Example Experiment: Learning from Examples

4.1   What Is SPARK by Example?

When wanting to learn to prove C programs using Frama-C and the WP plugin, *ACSL by Example* [7] is a good entry point. *ACSL by Example* is a booklet presenting how to prove with Frama-C more than 50 algorithms extracted from the C++ *Standard Template Library* (STL) [18,19]. Each algorithm in *ACSL by Example* comes with detailed explanations on how to specify it, how to prove it and what possibly are the lemmas needed for the proof. It is therefore a good companion for who wants to learn deductive methods with Frama-C and ACSL.

Even if some interactive learning tools for SPARK have been developed recently [1,2]) and good learning material is already available [3,15], we thought that a "recipe" document in the spirit of *ACSL by Example* was lacking for the SPARK community. Léo Creuse and Joffrey Huget were two second year students willing to learn formal methods during their S4 at ISAE-SUPAERO. We asked them to produce *SPARK by Example*, a SPARK equivalent of *ACSL by Example*, with the following constraints:

- All algorithms proved in *ACSL by Example* should be specified, implemented and proved in SPARK. SPARK Community 2018, freely available, must be used.
- A complete documentation must be produced for each algorithm detailing the specification and the implementation of the algorithm as well as proof "tricks" used for difficult proofs.

- Wrong or incomplete specifications that sometimes seem natural for beginners should also be described in order to explain why they are wrong.
- All proofs must be done in the spirit of [10], i.e. only with automatic SMT solvers and no interactive proof assistants, whereas some proofs in *ACSL by Example* need Coq.

Léo and Joffrey managed to specify, implement, prove and document all algorithms in less than 3 months without previous knowledge of formal methods nor SPARK. *SPARK by Example* is available on [8] and is provided as a Github project with all documentation directly readable from the website. All code source, installation instructions and build artefacts are also provided.

4.2   A Corpus of Proved Algorithms

Algorithms are classified in *ACSL by Example* and therefore in *SPARK by Example* in different chapters.

- The first chapter deals with non mutating algorithms. These algorithms do not modify their input data. For instance, find returns the first index in an array where a value is located and count returns the number of occurrences of a value in an array.
  These algorithms are rather simple, but they serve as a starting point for the reader of *SPARK by Example*. They will thus be very important as they are used to present important information (how to define a contract, ghost functions, loop invariants, how to interpret counterexamples returned by provers etc).
- Chapter 2 deals with maxmin algorithms and is a particular subset of non mutating algorithms. Its algorithms simply return the maximum and minimum values of an array.
- Chapter 3 is about binary search algorithms. These algorithms work on sorted arrays and therefore have a temporal complexity of $\mathcal{O}(\log n)$. The classic binary−search is presented in this chapter.
- Chapter 4 deals with mutating algorithms, i.e. algorithms that modify their input data. They all are procedures that rotate, copy, or modify the order of elements in arrays to match properties. The algorithms in this chapter generally have two implementations: the first one is usually easier, because the content of the input array is copied in another array; the second implementation is done on the array itself and has sometimes lead to difficulties in the proof process.
  In the previous chapters, there has been no difference between the algorithm specification and implementation in C/ACSL or SPARK, but due to the availability of "real" arrays in SPARK, this is the first chapter in which important differences between *ACSL by Example* and *SPARK by Example* appear.
- Chapter 5 on numeric algorithms is a special chapter because it mainly focuses on overflow errors. For instance, when returning the sum of the values in an array or the scalar product of two arrays, overflow errors may occur,

particularly if the values are integer ones. Moreover, even if the final result is in the right range, the intermediate results can overflow and lead to an error. It is the only chapter that deals with these kinds of errors so it is a little bit besides the others.

- Chapter 6 focuses on one particular data structure, namely the binary heap. It presents a concrete implementation of the classic heap structure as a record consisting of a fixed-sized array and a size attribute. It implements the basic algorithms dealing with heaps (push_heap, pop_heap) but also other algorithms such as make_heap that returns a heap created from an input array, or sort_heap that returns a heap of size 0 but with a sorted list inside of it.
- Chapter 7 deals with sorting algorithms and is a short chapter: is_sorted checks whether an array is sorted or not, and partial_sort partially sorts (with a specific definition) an array.
- Finally, Chapter 8 presents three classic sorting algorithms: selection_sort , insertion_sort and heap_sort.

SPARK by Example is therefore a rather complete document to learn how to use SPARK to prove classic algorithms. Chapters have an increasing difficulty and we hope that the explanations we have written for difficult notions or proofs are sufficiently clear to help beginners.

4.3 Lessons Learnt

What have we learnt from this experiment? First, it is possible to prove relatively complex algorithms without previous knowledge of formal methods in a relatively small amount of time. Of course, this has to be put in perspective as our students can rely on *ACSL by Example* to understand how to prove most of the algorithms. On the other hand, they had to understand how to build complex proof with lemma functions in SPARK which was clearly not the approach chosen by *ACSL by Example*.

Second, we encounter two difficult points when proving complex algorithms:

- The proof of some verification conditions often requires reasoning on properties that can not be directly handled by SMT or automatic solvers, for instance inductive properties. In this case, the proof may be achieved with a proof assistant like Coq or discharged by an automatic solver guided by *lemmas*. Lemmas are mathematical theorems, possibly with hypotheses, that must be added to the theories available to the SMT solvers to prove the verification condition. Of course, lemmas must also be proved, either using a proof assistant or an automatic prover.

  In SPARK, there is no proper construction of lemmas as in ACSL. To work around this limitation, the user has to define a procedure and use contract-based programming to write the lemma: hypotheses of the lemma are the preconditions of the procedure, conclusions of the lemma are its postconditions. This "emulation" requires them to be instantiated within the code to prove, whereas lemmas in Frama-C are automatically used by the provers when necessary. The main advantage of the SPARK approach is the fact that the user can help the solver to prove the lemma using an actual implementation of the procedure, whereas some lemmas in *ACSL by Example* have to be proven with Coq when the SMT solvers fail to prove them.

Therefore, the prover of a program must understand which lemmas are needed for his or her proofs, prove the lemmas and manually add instances of the lemmas in the code in order to help the provers. This is quite equivalent to give a sketch of proof to the provers. Fortunately, certain forms of "templates" appear when implementing lemmas for proving them, so it becomes easier and easier. Moreover, the GNATprove tool is of great help when inserting lemmas in the code to be proved to understand where to place them.

– Auto-active proof of a program is a kind of proof in which the verification conditions (VC) generated from the specification and assertions of the program are discharged only by automatic provers. Most of the time, SMT solvers like Alt-Ergo, CVC4 or Z3 are used as they embed first-order theories that are suitable for program verification (bitvectors or arrays theories for instance). However, SMT solvers are limited. For instance, universal quantifiers in formulas are not handled in a complete way. Therefore, when a SMT solver cannot discharge a VC, there may two several explanations:

  - The VC is effectively false
  - The VC should be discharged but is not due to solver limitation

The second case may be difficult to understand and solve by beginners, because they have to understand for instance that SMT solvers instantiate formulas involving universal quantifiers using instantiation patterns called triggers. Therefore, when trying to prove a VC with universal quantification, adding a new trigger through may ease the solvers task. We face this problem in a first version of *SPARK by Example* where we had to split a complex assertion on arrays involving nested quantifiers using several auxiliary functions that were used as triggers for the solvers. Notice that SMT solvers are improving quickly and that some VC that cannot be discharged by a solver may be discharged easily by a future version.

5 Conclusion

We presented in this paper two teaching experiments on deductive methods for program verification. The first one takes the form of a classic course on formal methods, but in a very small amount of time, i.e. 17 h. We showed that it is possible for beginners to use tools like Frama-C/WP or GNATProve to verify imperative programs written in C or SPARK. Experience shows also that bottom-up approaches, in which students are using the tools from the beginning of the course and prove more and more complex properties are better than top-down approaches often used in the French "Grandes Écoles" system in which theory is first exposed before practising. The second experiment pretended the production of teaching material, namely *SPARK by Example*, to evaluate students capacity to learn deductive methods. It showed that it is possible for non-experts to use

SPARK to prove relatively complex programs, as an implementation of binary heaps was entirely proved using only SMT solvers. We hope that *SPARK by Example* will also be a useful tool to learn SPARK and deductive methods.

Due to the structure of ISAE-SUPAERO engineering program, students sometimes lack knowledge or background that is useful to specify programs or fully understand how the tools work. For instance, writing complex specifications requires some understanding on how memory is represented in the tools or knowledge of programming language semantics that is currently not taught at ISAE-SUPAERO. Understanding how SMT solvers work and why they may fail to prove a verification condition is also important to handle complex proofs.

Concerning the S5 lecture, industrial feedback given on the last session of the course is really important for the students. They understand that these techniques, although rather mathematical, are used in industry, particularly in aerospace. Qualification and certification are also addressed during the session to show students that a technique or a tool, however attractive or powerful it is, must be incorporated in a global process.

Some ideas arise from these experiences:

- First, we may begin with a more suitable language for deductive verification. In particular, C is a difficult language to use, mainly due to pointers arithmetics, even if ACSL is a really nice specification language to manipulate. WhyML and the Why3 platform [11,20] seem to be natural candidates for a first initiation to deductive verification.
- Create a S4 optional course on reliable software using SPARK. This course would be based on the bottom-up approach described previously and incorporate theoretical sessions when necessary to help students. For instance, a 3-h lecture and practical session on how SMT solvers work would be useful.
- Add more formal methods during the curriculum, particularly in the Critical Systems major. For instance, TLA+ [17] could be introduced in the distributed systems course to show students that formal methods can be used for distributed algorithms.

References

1. AdaCore. Advanced SPARK - online course (2018). https://learn.adacore.com/courses/advanced-spark/index.html
2. AdaCore. Introduction to SPARK - online course (2018). https://learn.adacore.com/courses/intro-to-spark/index.html
3. AdaCore and Altran UK Ltd. SPARK 2014's User Guide (2018). http://docs.adacore.com/spark2014-docs/html/ug/index.html
4. AdaCore and Altran UK Ltd. SPARK 2014's User Guide (2019). http://docs.adacore.com/spark2014-docs/html/ug/index.html
5. AdaCore and Thales. Implementation Guidance for the Adoption of SPARK (2018). https://www.adacore.com/books/implementation-guidance-spark
6. Baudin, P., et al.: ACSL: ANSI/ISO C specification language (2018). https://frama-c.com/download/acsl-implementation-Chlorine-20180501.pdf

7. Burghardt, J., Gerlach, J.: ACSL by Example (2019). https://github.com/fraunhoferfokus/acsl-by-example
8. Creuse, L. et al.: SPARK by Example (2018). https://github.com/tofgarion/spark-by-example
9. Dijkstra, E.W.: Guarded commands, nondeterminacy and formal derivation of program. Commun. ACM **18**(8), 453–457 (1975)
10. Dross, C., Moy, Y.: Auto-active proof of red-black trees in SPARK. In: Barrett, C., Davies, M., Kahsai, T. (eds.) NFM 2017. LNCS, vol. 10227, pp. 68–83. Springer, Cham (2017). https://doi.org/10.1007/978-3-319-57288-8_5
11. Filliâtre, J.-C., Paskevich, A.: Why3 — where programs meet provers. In: Felleisen, M., Gardner, P. (eds.) ESOP 2013. LNCS, vol. 7792, pp. 125–128. Springer, Heidelberg (2013). https://doi.org/10.1007/978-3-642-37036-6_8
12. Floyd, R.W.: Assigning meanings to programs. In: Schwartz, J.T. (eds.) Mathematical Aspects of Computer Science. American Mathematical Society, pp. 19–32 (1967) ISBN: 0821867288
13. Hoare, C.A.R.: An axiomatic basis for computer programming. Commun. ACM **12**(10), 576–580 (1969)
14. ISAE-SUPAERO. The ISAE-SUPAERO engineering program (2019). https://www.isae-supaero.fr/en/academics/ingenieur-isae-supaero-msc/engineering-program/
15. McCormick, J.W., Chapin, P.C.: Building High Integrity Applications with SPARK. Cambridge University Press, Cambridge (2015)
16. Kirchner, F., et al.: Frama-C: a software analysis perspective. Formal Asp. Comput. **27**(3), 573–609 (2015). https://doi.org/10.1007/s00165-014-0326-7
17. Lamport, L.: Specifying Systems: The TLA+ Language and Tools forHardware and Software Engineers. Addison-Wesley Professional, Boston (2002)
18. Plauger, P.J., et al.: C++ Standard Template Library. Prentice Hall PTR, New Jersey (2000)
19. International Organization for Standardization (2011), ISO/IEC 14882:2011
20. The Toccata team. Why3. Where programs meet provers (2018). http://why3.lri.fr/

Using Krakatoa for Teaching Formal Verification of Java Programs

Jose Divasón and Ana Romero[✉]

University of La Rioja, Logroño, Spain
{jose.divason,ana.romero}@unirioja.es

Abstract. In this work, we present a study of different support tools to teach formal verification of Java programs and show our experience with Krakatoa, an automatic theorem prover based on Hoare logic which allows students to interactively visualize the different steps required to prove the correctness of a program, to think about the used reasoning and to understand the importance of verification of algorithms to improve the reliability of our programs.

Keywords: Formal methods · Hoare logic · Automated theorem provers · JML · Krakatoa

1 Introduction

Formal verification of algorithms is a technique to ensure that a program is *correct* even before being implemented in some programming language, verifying that the program *does what it is supposed to do* for all the possible input data without the necessity of applying testing. Although formal verification is not (yet) a mandatory subject in most computer science studies, it is included in the *Common Criteria for Information Technology Security Evaluation* [1].

In the University of La Rioja, formal verification is taught as part of a compulsory course called "Specification and Development of Software Systems" (SDSS). In the first years of existence of this course (the degree in computer science is taught in our University since 2002), formal verification was considered only in a "theoretical" way, explaining the Hoare logic axioms [12] and presenting the inference rules that make it possible to prove that a program satisfies a specification (given by means of a precondition and a postcondition). Six courses ago we decided to complement this teaching by means of some support tool to formally verify Java programs in a semi-automatic way. To this aim, we did a study of the available software for this task (some of them used only in companies or research) and the chosen support tool was Krakatoa [8]. In the first year of use of this theorem prover, we only used it as support tool during theoretical lessons, showing to students some basic examples. Then, we tried to improve the experience and since 2014 we decided to include in the course some practical lectures in a computer classroom where students could use the tool themselves.

© Springer Nature Switzerland AG 2019
B. Dongol et al. (Eds.): FMTea 2019, LNCS 11758, pp. 37–51, 2019.
https://doi.org/10.1007/978-3-030-32441-4_3

The paper is organized as follows. In Sect. 2 we present the context of the course SDSS and the teaching of formal verification in our university. Next, in Sect. 3 we study several alternatives for teaching formal verification of Java programs, considering different criteria which are interesting for our lectures, and we explain why Krakatoa is the most adequate for our purposes. Some examples of formal verification with this tool are shown in Sect. 4 as well as the set of exercises that the students must solve and the question on verification in the final exam. We present the results of our experience in Sect. 5. Finally, conclusions and further work are detailed in Sect. 6.

2   Context of Our Experience

At the computer science studies at the University of La Rioja, formal verification is taught as part of the course "Specification and Development of Software Systems" (SDSS). SDSS is a compulsory course that is taught in the fourth semester of the degree in computer science, and corresponds to the fourth course of Programming. The course has 6 ECTS, divided into 30 h of theoretical lectures, 28 h of practical exercises in a computer laboratory, 2 h for the final exam and 90 h of the student individual work. As it is claimed in the guide of the course, one of the aims of SDSS is to provide a formal perspective about different aspects of programming (syntax, semantics, correctness and efficiency), trying to improve the programming skills of students. After considering some subjects such as specification and implementation of abstract data types and their relation with object-oriented programming and specification of algorithms, the final part of the course (about 15 h) is devoted to formal verification of algorithms. The course SDSS is also part of the degree in mathematics; the students of both degrees attend together the lectures and the contents and the evaluation system for all of them are the same. The students are supposed to have followed the three previous courses on Programming. Moreover, they are supposed to have acquired the fundamental concepts of first order logic which are taught in the second semester of the degrees (as part of the course "Logic"). Each year there are around 70 students, of which about two thirds are students of computer science the rest of mathematics. With respect to the evaluation, 70% of the mark corresponds to the final exam and 30% to laboratory exercises.

One of the goals of SDSS is to consolidate the acquired knowledge in the third semester of both degrees, where it is introduced the concept of object-oriented programming in Java. For this reason, Java is the chosen programming language for the SDSS course. It is worth noting that at the beginning of the course, the students lack a strong experience in programming languages. This, in conjunction with the few hours available to the formal verification part, cause that the introduction from scratch of another different programming language could be a counter-productive decision.

Until 2013, formal verification was taught in SDSS only in a "theoretical" way by means of the Hoare logic axioms [12]. Given a precondition Q and a postcondition R, a program "s" (consisting of a sequence of elementary instructions $s \equiv \{s_1, \ldots, s_n\}$) satisfies the specification $\{Q\}s\{R\}$ if: whenever the program s is executed starting in a state which satisfies Q, the program terminates and the final state satisfies R. In order to verify the correctness of $\{Q\}s\{R\}$, one must consider predicates which determine the states which are satisfied at the intermediate points of the program, called *assertions*, such that $\{Q\}s_1\{P_1\}s_2\{P_2\}\ldots\{P_{n-1}\}s_n\{R\}$. If the initial assertion Q (precondition) is satisfied, and each elementary "program" s_k, consisting of one simple instruction, satisfies the specification $\{P_{k-1}\}s_k\{P_k\}$, then when the program stops the postcondition R is satisfied and therefore the program is correct. In our course, we do not deal with *partial correctness* but with *total correctness*, i.e., a program is correct when it returns the expected result and the algorithm terminates.

Hoare logic provides rules to verify the correctness of the elementary instructions of a programming language (assignments, sequential composition, conditional clauses and iterative composition). These rules allow one to compute in a straightforward way correct preconditions, from a given postcondition, for the cases of assignments, sequential compositions and conditional clauses. However, in the case of the iterative composition the process is not direct and it is necessary to construct first an *invariant* predicate P and a *variant* V. Then, Hoare logic requires that the loop body decreases the variant (to ensure termination) while maintaining the invariant. In addition, the invariant must be strong enough so that at the end of the loop we could deduce the postcondition. Usually, the students find it difficult to figure out the invariant.

In SDSS we present (in a theoretical way) the Hoare rules for the basic instructions of an iterative language and we do small examples of application of each one of the rules. Once all the rules have been introduced, we do some exercises of formal verification proofs of some small programs with an iterative scheme. The proofs of correctness considered in the course SDSS are *restricted* to programs corresponding to the following sketch:

```
{Q}
<init>
while B do {
    <body>
}
<end>
return <var>
{R}
```

where the blocks <init>, <body> and <end> consist of a sequence of elementary instructions, usually assignments and conditional structures. In fact this is not a restriction, because if there are several "sibling" loops it can be thought that all but the last one are inside <init>, and if there are nested loops one can think that the internal loops are inside <body>.

Taking into account Hoare's axioms, in order to verify the correctness of a program with the previous sketch it is necessary to:

1. Find an invariant P for the loop.
2. Verify the specification $\{Q\}$<init>$\{P\}$.

3. Verify that P is an invariant, that is to say, the specification $\{P \text{ and } B\}$<body>$\{P\}$ is satisfied.
4. Verify the specification $\{P \text{ and } \text{not}(B)\}$<end>$\{R\}$.
5. Find a variant.

Following these steps, in SDSS we consider proofs of correctness of some easy algorithms such as the (iterative) computation of the power of a real number raised to a natural number, the computation of the factorial of an integer, the integer square root, the sequential search of an element in an array and the sum of all components of an array. After explaining some of these exercises on the blackboard, we do also some exercise classes where the students must apply their knowledge in a practical way and make some correctness proofs on their own. The difficult part of the exercises is the determination of the invariant P, and it is very frequent that students propose invariants that are not strong enough and they have to make different attempts (and repeat steps 2, 3 and 4 for all of them) in order to find the correct one.

This "traditional" way of teaching formal verification was the chosen one at our University until 2013. At that moment, we decided to complement the theoretical lectures with the help of some support automatic tool for formal verification of Java programs based on Hoare logic. A study of the different alternatives was done, and the chosen tool was Krakatoa, an automatic theorem prover which allows students to interactively visualize the various steps required to prove the correctness of a Java program, to think about the used reasoning and to understand the importance of verification of algorithms to improve the reliability of our programs. In the first year of use of this theorem prover, we only used it as support tool during theoretical lessons, showing to students some basic examples of formal proofs with Krakatoa. After the positive results of that initial attempt (the marks in the final exam of the formal verification part were higher than previous years and students showed interest in Krakatoa), we tried to improve the experience and since 2014 we decided to include in the course some practical lectures in a computer classroom where students could use the tool themselves, providing the correct specification and the necessary code to verify the proposed programs as explained in Sect. 4.

3   Study of Different Alternatives

Our study of different alternatives for teaching formal verification of Java programs started in 2013, when we decided to complement our theoretical lessons with some support tool. At that moment we found some documentation of universities where formal verification was taught in a practical way (see for example [7,14,15]), but most of them did not correspond to Java programs or did not seem to be based on Hoare logic.

There are several approaches and levels to carry out formal verification of programs. Essentially, tools for this task are classified in three groups. Interactive theorem provers, such as Isabelle [18] and Coq [6], belong to this kind of tools.

They allow a mathematical modeling and verification of programs at the highest level of confidence (Common Criteria certification at level EAL7). They have been used in industrial applications, such as the verification of seL4, an operating system kernel [13]. Nevertheless, these tools need a steep learning curve to gain enough expertise to be able to prove formally specifications of programs, so they seemed not to be a good choice for a first introduction course on formal verification. Secondly, there are tools based on model checking, such as Java Pathfinder [11]. This kind of tools are supposed to be a rigorous method to find a violation of a given specification, not only by means of tests but with abstract interpretations. Finally, there exist tools based on Hoare logic or similar logics, which are the ones we were mainly interested in due to their relation with the theoretical part of the course. In addition, some such tools are indeed focused on teaching (but not in Java), such as Dafny [14] and HAHA [17]. Due to the context of SDSS, we aim to use a tool devoted to verify Java programs. Then, we selected the following tools for evaluation.

- Krakatoa [8] (version 2.41, May 2018)
- KeY [2] (version 2.6.3, October 2017)
- OpenJML [5] (version 0.8.40, October 2018)

Those ones seem to be the most important ones, although there exist more alternatives that could also have been considered for this study such as Jahob [4] and Jack [3]. However, most of them are no longer developed.

In the concrete context of our SDSS laboratory lessons, we evaluated the following features of the tools: ease of use (taking into account that the program will be used by students with no previous knowledge on it), feedback (the information about the proof attempts and proof failures should be understandable for students with no expertise in the tool), documentation (the evaluated tool should have enough examples of different levels of difficulty), relation between the tool and the contents that are taught in the theoretical lessons (this is the most important feature for us: we want to check if the tool clearly follows the steps from Hoare logic), ease of installation and if there exist plugins for Eclipse (the IDE used in our laboratories) or an online tool. Table 1 shows a summary of this evaluation. Apart from that, we also checked the tools against seven exercises that we teach in SDSS in a theoretical way:

1. Minimum of two integers
2. Swap two elements of an integer array
3. Square root (linear version)
4. Square root (binary version)
5. Check if an integer array is sorted in ascending order
6. Exponentiation
7. Linear search of an element in an integer array

We checked if the language of each tool is expressive enough to specify the algorithms, and also whether the evaluated tools are able to prove them strictly by means of just the specification (precondition and postcondition), together with the corresponding invariants and a measure that decreases in each step if necessary. The result is shown in Table 2. It is worth remarking that most of the programs are able to verify automatically the exercises once the user has provided some hints (or working a bit with the corresponding goals), but we wanted to test them exactly with the same reasoning that we would make in the theoretical lessons: that is, just making use of the specification, invariants and variants (something that decreases in each step).

KeY is the most powerful tool, from the ones that we have studied, and it is also the most used one. It is worth noting that there are several universities where KeY is used as a tool for teaching formal verification of Java programs such as Chalmers University[1] or the Karlsruhe Institute of Technology[2], but within the computer science master's programme. In our opinion, KeY requires a longer learning step than Krakatoa and it is to be used by experts, or at least, it is not designed to be used by degree students in the forth semester. In addition, KeY was not able to automatically prove the correctness of our examples (just from the specifications, invariants and variants), but Krakatoa had a higher success rate. To sum up, despite of the fact that Krakatoa is not the most powerful one, it fits our requirements. Thus, we decided to put it into practice in our laboratory lessons. The Krakatoa program was then available in the computer laboratories of our university as an Eclipse plugin. Indeed, it is a virtualized application, i.e., the students can use it at home easily. This solves the two main drawbacks which Krakatoa presents in our study: the lack of an online tool and the difficulty of its installation. The confusing feedback provided by Krakatoa is solved with the help of the teachers in the laboratory lessons.

It is worth noting that the study of these tools was repeated every year from 2013 (we present here the one that we did in January 2019). The performance of KeY with some exercises has improved in the last years but the results of all studies were similar.

Table 1. Main features of the evaluated tools.

	Tool		
	Krakatoa	KeY	OpenJML
Ease of use	✓	✗	✓
Feedback	Lack of information	Need a deep knowledge	✓
Related to theory	✓	✓	✗
Documentation	Few examples	✓	Under development
Ease installation	✗	✓	✓
Plugin Eclipse	✓	✓	✓
Online tool	✗	✓	✓

---

[1] http://www.cse.chalmers.se/edu/year/2018/course/TDA294_Formal_Methods_for_Software_Development/.

[2] https://formal.iti.kit.edu/teaching/FormSys2SoSe2017/.

Table 2. Expressiveness and solvable problems by the tools.

	Tool					
	Krakatoa		KeY		OpenJML	
	Specif.	Solv.	Specif.	Solv.	Specif.	Solv.
Minimum	✓	✓	✓	✓	✓	✓
Swap two elements	✓	✓	✓	✓	✓	✓
Linear sqrt	✓	✓	✓	✗	✗	✗
Binary sqrt	✓	✗	✓	✗	✗	✗
Sorted array	✓	✓	✓	✓	✓	✓
Exponential	✓	✓	✓	✗	✗	✗
Linear search	✓	✓	✓	✗	✓	✓

4 Some Examples of Formal Verification with Krakatoa

Trying to complement the theoretical teaching of formal verification by means of some software, and after the study of alternatives explained in Sect. 3, the chosen support tool has been Krakatoa. As already said in Sect. 2, in an initial experience we used it only as a support tool during theoretical lessons but since 2014 we decided to include in the course some practical lectures in a computer classroom where students could use the tool. More concretely, we have now 3 practical lectures (each of them of 2 h) for 3 different levels of exercises. The first lecture is devoted to the specification and verification of Java methods where only assignments and conditional clauses are used; in this case, if the specification given by the student is valid, Krakatoa should be able to verify directly that the program is correct. On the contrary, if iterative structures are included, Krakatoa needs some *help* and the student must write the invariant predicate for the loops; the second lecture is devoted to this kind of exercises. Finally, it is also sometimes necessary to introduce auxiliary *predicates, axiomatic definitions* or *assertions*, which are explained in the third lecture. Since formal verification is only part of the course contents, we teach it in an introductory way and we do not have time to teach formal verification of object oriented aspects such as classes, inheritance or dynamic types.

In this section, we present some examples that we show in the lectures, the mandatory exercises that the students must solve in the computer laboratory and the verification exercises of the final exam.

4.1 Lectures in the Computer Classroom

As we have already said, we have now 3 practical lectures for 3 different levels of exercises. We present here one example of the exercises explained in each one of the lectures. Other examples of exercises of formal verification of Java programs explained in SDSS can be found in [16].

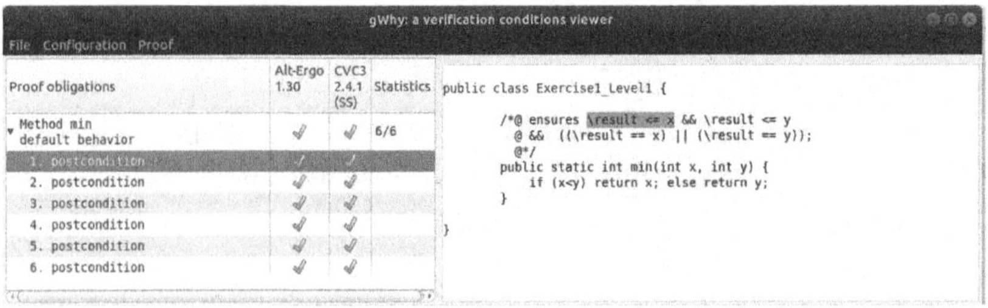

Fig. 1. Obligations generated by Krakatoa for the method min.

In order to verify the correctness of a Java program, Krakatoa inputs the specification (precondition and postcondition) written in the Java Modeling Language [10] (JML). Then, making use of a tool called Why [9], it generates a series of lemmas (called *proof obligations*) that correspond to the different steps, following Hoare logic, to verify the correctness of the program. These lemmas must be verified by some automatic theorem provers which are included in the Krakatoa tool; if these theorem provers do not reach some of the proofs, it is also possible to send the lemmas to Isabelle and Coq, two interactive theorem provers where the user can help the prover to construct the proofs.

Minimum of 2 Elements. One of the simplest examples explained in the first Krakatoa practical lesson is the following Java method for computing the minimum of two integers x and y:

```
/*@ ensures \result <= x && \result <= y &&
  @ ((\result == x) || (\result == y));
  @*/
public static int min(int x, int y) {
    if (x<y) return x; else return y;
}
```

The specification of the method, in JML, is written as a comment between `/*@` and `@*/`. The clause `ensures` is used to introduce the postcondition, which is a logical predicate which must be satisfied when the method stops for any possible value of the inputs. Inside the postcondition, `result` is used to denote the returned value. In this case, the postcondition means: the result is smaller than or equal to x, the result is smaller than or equal to y, and the result is equal to x or equal to y.

The goal of Krakatoa consists of verifying that the method `min` is implemented in a correct way, that is to say, it satisfies the given specification. As shown in Fig. 1, Krakatoa generates 6 lemmas (proof obligations) that express the correctness of the program. The 6 obligations correspond to each one of the 3 components of the postcondition, which must be satisfied by each one of the two branches of the conditional clause. These obligations are the steps that the students should do to formally verify (in a theoretical way) that the program is correct. The first lemma, which is detailed on the right side of Fig. 1, says that, the result is less than or equal to x. In this example the lemmas are very easy and the automatic theorem provers Alt-Ergo[3] and CVC3[4], which are integrated in Krakatoa, are able to verify them in a direct way. The proof of the 6 obligations imply that the program is correct with respect to the given specification, which ensures that in any possible situation, that is to say, for any of the infinite possible input data, the method returns the desired result.

Deciding if an Array Is Sorted. The correctness proof of a program is more complicated when it includes iterative instructions. Let us consider now the following method to decide if the elements of an array of integers are sorted (in ascending order):

```
/*@ requires v != null && 1 <= v.length ;
  @ ensures \result <==> (\forall integer j; (0 <=j < v.length-1) ==>
  @ v[j]<=v[j+1]);
  @*/
static boolean isSorted(int v[]) {
    int i=0; boolean b=true;
    while (i<v.length-1 && b) {
        if (v[i] > v[i+1]) b=false;
        i=i+1;
    }
    return b;
}
```

The clause **requires** introduces the precondition, which is a logic predicate that must be satisfied when the method is called. In this case, the argument v must be a non-null array with positive length. Krakatoa generates now 2 obligations corresponding to the postcondition but, as one can observe in Fig. 2, it is not able to prove them. From the 8 obligations which ensure that the method is safe, it only proves 5 of them.

In order to verify the correctness of a Java program with iterative instructions following the axioms of Hoare logic, as we have explained in Sect. 2, it is necessary to define an invariant P which is a predicate that is satisfied at the beginning and end of each execution of the loop. This invariant must be strong enough so that when the loop finalizes the postcondition is satisfied. In general it is a difficult problem to find the adequate invariant.

In order to introduce the invariant predicate in the JML specification of a program in Krakatoa one uses the clause **loop invariant**. Moreover, to be able to verify that the loop stops (and therefore the method is safe), very frequently we must define in Krakatoa the variant, which must be an integer expression such that it is non negative and it decreases after each execution, denoted by **loop variant**. For the iterative structure inside the method **isSorted** we can use the following specification:

---

[3] Alt-Ergo http://alt-ergo.lri.fr/.

[4] CVC3 http://www.cs.nyu.edu/acsys/cvc3/.

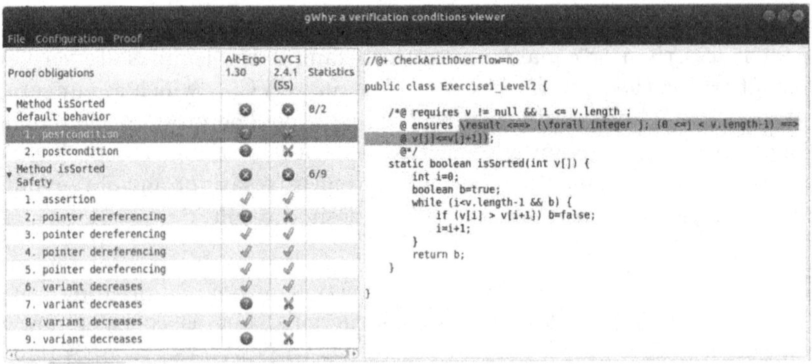

Fig. 2. Obligations generated by Krakatoa for the method isSorted without specifying the invariant predicate.

```
static boolean isSorted(int v[]) {
    int i=0; boolean b=true;
    /*@ loop_invariant 0<=i && i<v.length && (b == true <==>
      @ (\forall integer j; (0 <=j < i) ==> v[j]<=v[j+1]));
      @ loop_variant v.length-i;
      @*/
    while (i<v.length-1 && b) {
        if (v[i] > v[i+1]) b=false;
        i=i+1;
    }
    return b;
}
```

Krakatoa generates now 21 obligations, some of them have appeared when the invariant has been introduced. The proof of such obligations will show the soundness of the algorithm. We can also observe than the generated obligations correspond to steps of the *theoretical* proof of the correctness of the program explained in Sect. 2. With the help of this invariant the Alt-Ergo and CVC3 theorem provers are able to verify the correctness of the program, as shown in Fig. 3.

Exponential Function. In some situations, the definition of the invariant predicate and the *variant* is not enough to prove the correctness of a program with iterative structures and it is also necessary to include auxiliary predicates, axiomatic definitions and *assertions* which help the theorem provers to verify the lemmas generated by Krakatoa.

The following method implements the exponential function raising a float to an integer:

```
public static float exponential (float x, int n) {
    int i=0; float r=1;
```

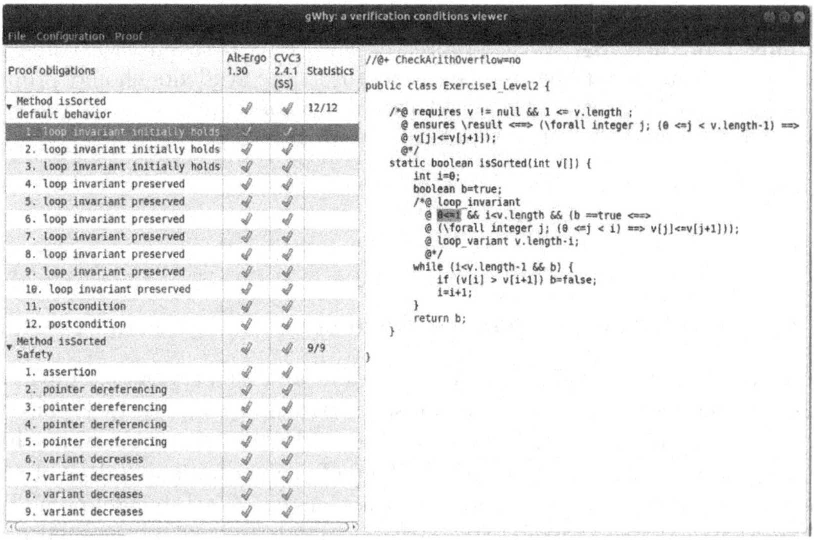

Fig. 3. Obligations generated by Krakatoa for the method isSorted after specifying the invariant predicate.

```
    while (i<n) {
        i++;
        r=r*x;
    }
    return r;
}
```

After introducing the idea of the method, the students must think of a possible specification written in JML. Since the JML version supported by Krakatoa does not allow to use the exponential function, in order to specify the method it is necessary to include the following axiomatic definition:

```
/*@ axiomatic Exponential {
  @ logic float exp(float x, integer n);
  @ axiom exp_zero : \forall float x; exp(x,0) ==1;
  @ axiom exp_sum: \forall integer n; \forall float x;
  @ exp(x,n+1) == exp(x,n)*x;
  @}
@*/
```

Using this axiomatic definition, the students should write, using the JML syntax, the specification of the method `exponential`:

```
/*@ requires n >=0;
  @ ensures \result == exp(x,n);
  @*/
```

With this specification, Krakatoa generates 3 obligations but it is able to prove only 1 of them. As we have already said, it is necessary to define the invariant predicate P for the iterative structure. The students should propose an invariant, run 'Krakatoa and see if the obligations are proved. A possible solution for the invariant (and variant) of the program is:

```
/*@ loop_invariant 0 <= i && i <= n && r == exp(x,i);
  @ loop_variant n-i;
  @*/
```

4.2   Exercises in the Computer Classroom

Once Hoare logic, the steps for verifying programs and the previous examples are explained in the lectures, the students must practice and complete by themselves some exercises with Krakatoa. To this aim, the students work in pairs. They have 6 h at the computer laboratories with the help of the teacher and 2 days of work at home before the deadline to send their solutions via GitHub classroom. This set of exercises consists of two parts and corresponds to a 5% in the final mark of the course. The first part is mandatory and has 8 exercises devoted to design and verify the following Java programs:

1. A method to compute the absolute value of an integer
2. Check if the arithmetic mean of three non-negative real numbers is higher or equal to 5
3. A method to compute the maximum of three integer numbers
4. Given an array with 4 real numbers, modify it by dividing each component by the sum of all components (with no loops)
5. Decide if a number is prime
6. Check if all elements in an integer array are non-negative
7. Compute the highest factor of a positive integer number (excluding itself)
8. A method to compute the factorial of a non-negative integer number

The second part comprises three voluntary exercises: modification of each component of an integer array by its absolute value, find the frequency of a number in an array and finally design and verify other algorithms with loops or that use some of the previous exercises.

In the course 2019, 61 students (from 68) did the set of Krakatoa exercises. As in previous years, they had very good marks: the mean was 0.457 (over 0.5). Concretely, the students of computer science had a mean of 0.447 and the ones of mathematics 0.467. Table 3 shows the number of students with wrong answers in each mandatory exercise (all students did all of them). As it can be seen, the students have problems with exercise 7, which corresponds to the one with the most difficult invariant. With respect to the optional exercises, 80% of the students did the first one (from which, 90% did it well). Only 26.9% of the students sent the second voluntary exercise (all of the received answers were correct). One student did the third one.

Table 3. Results of mandatory exercises in the computer classroom. Number of students who did the exercises N = 61.

Exercise	1	2	3	4	5	6	7	8
Wrong answers	2	1	4	3	2	2	10	4

4.3   The Exam

The final exam consists of three written exercises of the different aspects covered by the course. The exercises are solved without the help of the computer. One of them is about formal verification. It is the most important one: approximately 45% of the mark in the final exam corresponds to this exercise.

In the last course (2019), this exercise consists of two parts:

1. Prove that the predicate $P = (1 \leq i \leq n)$ and $(n \% m = 0)$ and $(\forall \alpha \in \{m+1, \ldots, i-1\}.(n \% \alpha \neq 0))$ is an invariant for the following loop:

```
while (i < n) {
    if (n % i == 0) m=i;
    i++;
}
```

2. Verify the correctness of an algorithm computing the mean of the values of the elements of an array.

In general, the marks in this exercise were high (7.35 over 10 last year, the higher the better). During these years, we noticed a better understanding of the concepts among the students of the degree in mathematics, since they are more used to abstract reasoning and formal proofs. Indeed, the students of that degree outperform the students of the degree in computer science. This can be shown, for instance, in the marks of the verification exercise of the last exam: the mean of the marks of the students of computer science was 6.29, whereas the mean increased to 8.19 for the students of the degree in mathematics.

5   Results of the Experience

The results of using Krakatoa as a support tool for teaching formal verification of Java programs have been very positive. First of all, we have observed that after using Krakatoa the students understand the different steps of the (theoretical) formal proofs in a better way; more concretely, when Krakatoa was not used as a support tool many of the students *memorized* the exercises of formal verification (and very frequently they did not really understand them). This better understanding of students has been shown in the marks on average of the formal verification exercises in the final exam that have increased significantly (see Table 4, exercises are marked with a number between 0 and 10, the higher the better). The first year of use of Krakatoa just as a support tool (2013), the average of the marks in the final exam of the formal verification part was

Table 4. Marks on average of formal verification exercises in the final exam.

Year	Students	Marks
2012	46	6.14
2013	50	7.50
2014	37	7.06
2015	38	7.17
2016	54	6.81
2017	54	7.23
2018	73	7.86
2019	66	7.35

higher than previous years. The difficulty was very similar. During the following courses the marks remained higher than in 2012, although deeper contents and higher difficulty of the exercises were demanded in the exams. Moreover, many students claim now that this is the most interesting part of the course. Indeed, two students decided to carry out their final-degree project on this subject. Our experience as teachers is also positive and we plan to continue using Krakatoa in the following courses.

6   Conclusions and Further Work

In this work we have presented our experience with the tool Krakatoa to teach formal verification of Java programs, improving in this way the theoretical lessons on Hoare logic and helping students to understand the different steps of formal verification. With this experience, the average marks of formal verification exercises in the final exam has been increased; moreover, students show interest in this part of the course.

After these positive results, we plan to continue using Krakatoa in the following courses in laboratory sessions, considering other exercises with similar difficulty to the ones presented in this work. We will also repeat the study of other possible tools presented in Sect. 3, and we think that KeY could be also a good candidate in the future.

Acknowledgments. Partially supported by the Spanish Ministry of Science, Innovation and Universities, project MTM2017-88804-P.

References

1. Common Criteria for Information Technology Security Evaluation. Technical report (2012)
2. Ahrendt, W., Beckert, B., Bubel, R., Hähnle, R., Schmitt, P.H., Ulbrich, M. (eds.): Deductive Software Verification - The KeY Book - From Theory to Practice. Lecture Notes in Computer Science, vol. 10001. Springer, Berlin (2016). https://doi.org/10.1007/978-3-319-49812-6

3. Barthe, G., et al.: JACK — a tool for validation of security and behaviour of Java applications. In: de Boer, F.S., Bonsangue, M.M., Graf, S., de Roever, W.-P. (eds.) FMCO 2006. LNCS, vol. 4709, pp. 152–174. Springer, Heidelberg (2007). https://doi.org/10.1007/978-3-540-74792-5_7

4. Bouillaguet, C., Kuncak, V., Wies, T., Zee, K., Rinard, M.: Using first-order theorem provers in the Jahob data structure verification system. In: Cook, B., Podelski, A. (eds.) VMCAI 2007. LNCS, vol. 4349, pp. 74–88. Springer, Heidelberg (2007). https://doi.org/10.1007/978-3-540-69738-1_5

5. Cok, D.R.: OpenJML: JML for Java 7 by extending OpenJDK. In: Proceedings of the Third International Conference on NASA Formal Methods, pp. 472–479. NFM 2011 (2011)

6. Coq development team: The Coq Proof Assistant, version 8.9.1. Technical report (2019). https://coq.inria.fr/

7. Feinerer, I., Salzer, G.: Automated tools for teaching formal software verification. In: Proceedings of the 2006 Conference on Teaching Formal Methods: Practice and Experience, p. 4 (2006)

8. Filliâtre, J.C., Marché, C.: The Why/Krakatoa/Caduceus platform for deductive program verification. In: Proceedings of the 19th International Conference on Computer Aided Verification, pp. 173–177. CAV 2007 (2007)

9. Filliâtre, J.-C., Paskevich, A.: Why3 — where programs meet provers. In: Felleisen, M., Gardner, P. (eds.) ESOP 2013. LNCS, vol. 7792, pp. 125–128. Springer, Heidelberg (2013). https://doi.org/10.1007/978-3-642-37036-6_8

10. Gary T., Leavens, A.L.B., Ruby, C.: Preliminary design of JML: a behavioral interface specification language for Java. Technical report (2000), iowa State University

11. Havelund, K., Pressburger, T.: Model checking Java programs using Java PathFinder. Int. J. Softw. Tools Technol. Transf, 2(4), 366–381 (2000)

12. Hoare, C.A.R.: An axiomatic basis for computer programming. Commun. ACM 12, 576–580 (1969)

13. Klein, G., et al.: seL4: formal verification of an OS kernel. In: Proceedings of the ACM SIGOPS 22nd Symposium on Operating Systems Principles, pp. 207–220. ACM (2009)

14. Leino, K.R.M.: Dafny: an automatic program verifier for functional correctness. In: Clarke, E.M., Voronkov, A. (eds.) LPAR 2010. LNCS (LNAI), vol. 6355, pp. 348–370. Springer, Heidelberg (2010). https://doi.org/10.1007/978-3-642-17511-4_20

15. Poll, E.: Teaching program specification and verification using JML and ESC/Java2. In: Gibbons, J., Oliveira, J.N. (eds.) TFM 2009. LNCS, vol. 5846, pp. 92–104. Springer, Heidelberg (2009). https://doi.org/10.1007/978-3-642-04912-5_7

16. Romero, A.: El uso de los demostradores automáticos de teoremas para la enseñanza de la programación. In: Proceedings of Jornadas de Enseñanza Universitaria de la Informática (JENUI 2013) (2013)

17. Sznuk, T., Schubert, A.: Tool support for teaching Hoare logic. In: Giannakopoulou, D., Salaün, G. (eds.) SEFM 2014. LNCS, vol. 8702, pp. 332–346. Springer, Cham (2014). https://doi.org/10.1007/978-3-319-10431-7_27

18. T. Nipkow, M. Wenzel, L.C.P.: Isabelle 2019 (2019). https://isabelle.in.tum.de/

Teaching Deductive Verification in Why3 to Undergraduate Students

Sandrine Blazy$^{(\boxtimes)}$ (iD)

Univ Rennes, Inria, CNRS, IRISA, Rennes, France
sandrine.blazy@irisa.fr

Abstract. We present the contents of a new formal methods course taught to undergraduate students in their third year at the University of Rennes 1 in France. This course aims at initiating students to formal methods, using the Why3 platform for deductive verification. It exposes students to several techniques, ranging from testing specifications, designing loop invariants, building adequate data structures and their type invariants, to the use of ghost code. At the end of the course, most of the students were able to prove correct in an automated way non-trivial sorting algorithms, as well as standard recursive algorithms on binary search trees.

1 Introduction

Since the design by contract methodology implemented in Eiffel in the eighties [13], many programming languages support assertion-based contracts, that are taught in programming and software engineering courses for beginner students. As an example, in our university, contract-based programming is introduced at the end of a second-year software engineering course where students learn the Scala programming language [15].

Moreover, automated provers (especially SMT solvers) improved significantly over the decade. They are applied in many different fields, including formal verification, and are able to prove huge expressive formulae. Several tools are now available to automatically prove that a program is correct with respect to its specification. These deductive verification tools generate verification conditions (i.e., logical formulae) that are discharged by automated provers. Among them are Dafny [12], Frama-C [6], KeY [3], Rodin [2], SPARK [5], Verifast [10], Viper [14] and Why3 [7]. Some of them operate over real languages such as Ada, C and Java and are used in industry to develop real-world programs [5]. They rely on an expressive specification language (e.g., JML [11], or EACSL [4]).

Encouraged by these advances, we decided to teach deductive verification to undergraduate students in their third year at university. After 25 years of teaching various formal methods courses at master level using different tools, we found that it was the right time to teach deductive verification at the undergraduate level. Our class consists of about 100 students. Most of them followed in previous years introductory courses on functional and immutable programming (using

© Springer Nature Switzerland AG 2019
B. Dongol et al. (Eds.): FMTea 2019, LNCS 11758, pp. 52–66, 2019.
https://doi.org/10.1007/978-3-030-32441-4_4

Scala, first year), and on software engineering (second year). Other students are newcomers with a different background. During their third year and before our course, they follow an introductory course on logic and a programming course in Java on basic data structures.

The goal of our course is to learn how to formally specify a program, in order to prove that the instructions of the program satisfy its specification. An effort is done on studying various practical case studies, involving non-trivial algorithms, where students practice a lot with a tool, and use the tool to learn from their errors. Students learn what precise pre and postconditions are, they practice loop invariants and define type invariants to make their specifications more concise. Mastering these notions takes some time and writing precise loop invariants is still difficult for beginners. To guide the students in their formal development process, we require them to follow a pedagogical approach that we describe in this paper. The course briefly introduces the weakest-precondition calculus used to compute verification conditions, but it is not a course on proof techniques and interactive proofs. Such courses are taught the following year during the first year of master studies.

The tool we chose for our course is the Why3 platform [7]. Its programming language is an imperative programming language, that is the intermediate language used by different tools for verifying Ada, C and Java programs. In past teaching experiences, we tried other tools operating over a more complex language, but it required the students to understand precisely the semantics of tricky features of the language, and to frequently pollute their specifications with minor details about these features. Why3 supports a simple and clear programming language. Moreover, beginners make frequent and various mistakes when writing a specification or a piece of code, and they need to understand their errors. We found that Why3 is a very useful tool to help them find and correct their errors. Last, we also chose Why3 because of our geographical proximity with the Why3 development team.

Our course is organized as follows. Seven lectures (each one lasts two hours and so do exercise and lab sessions) present general ideas and concepts and live demonstrations. Each lecture is followed by one or two exercise sessions, where students practice in group setting. There are eight exercise sessions and each exercise session is followed by one or two lab sessions. The goal of an exercise session is to prepare the following lab session. Each exercise session ends with a small quiz. Each quiz gets a score; it is corrected and returned by email before the following lab session. These quizzes make the students more attentive during the lectures and the exercise sessions. A typical quiz question is a choice between four specifications of a program, where we reuse wrong answers from previous years to build these four specifications. Such various choices given to the students may generate interesting live discussions. There are ten lab sessions, where students work in pairs in small-group settings. We use the Moodle learning platform of our university to collect the student exercises done during lab sessions and to communicate with them. Last, the semester ends with a written exam.

The rest of the paper is organized as follows. First, Sect. 2 gives background information on the Why3 platform. Then, Sect. 3 highlights the first notion taught to students, precise specifications. It concerns pre and post-conditions, and loop invariants. Section 4 explains how we use type invariants and ghost code to write more complex programs, and Sect. 5 details some examples of recursive programs. Finally, Sect. 6 concludes.

2   Background on Why3

From a user point of view, Why3 provides a rich specification and programming language called WhyML [8]. The specification language of Why3 is an extension of first-order logic. The programming language of Why3 is an imperative language with arrays, polymorphism, algebraic data types, pattern matching and references. It also features exceptions, but we do not use them in our course (we restrict ourselves to option types to model a failure of a program). Its syntax is close to OCaml syntax. Why3 comes with a standard library providing useful definitions and properties of common data structures. We mainly use the following libraries: integers, maps, arrays, matrices, lists, sequences and trees.

The Why3 user writes a specification and instructions implementing this specification. Why3 translates them into verification conditions that are given as input to external automated provers. An interesting feature of Why3 is that it is able to translate and communicate the verification conditions to many external provers [7]. The next step is the automated proof that the program satisfies its specification. Why3 provides different strategies to conduct automatically this proof. In our installation, the basic strategy (called 0) mainly calls the SMT solver Alt-Ergo; other strategies transform formulae (via splitting of conjunctions and unfolding of definitions) before calling several automated solvers. Compared to strategy 1, in strategy 2, the timeout and memory limit of the external provers are increased. In our course, we only use the four standard strategies (called 0, 1, 2 and S, where S is the default strategy for splitting a conjunctive formula) and the four automated provers Alt-Ergo, CVC4, Eprover and Z3. Fortunately, we neither need to understand the internals of these strategies, nor to use ad hoc advanced interactive strategies provided by Why3.

When an automated prover manages to prove a verification condition, we know that it is valid (assuming that the tools are sound). When none of them is able to prove a verification condition, either there is an error in the instructions that do not satisfy this verification condition, or there is an error in the specification, or the verification condition holds but none of the automated provers is able to prove it. The first step to understand what is wrong is then to look at the logic formula that could not be proved. This formula is the verification condition expressed in WhyML. Let us note that Why3 can also be used as a tool to write and prove logical formulae. We do not use it as such, but we use a lot the logic window of its graphical user interface.

Figure 1 shows the graphical user interface of Why3. The upper right window is an editor to write the specification and the program. The example program

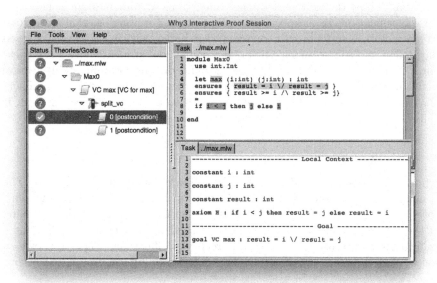

Fig. 1. The Why3 IDE. From top left to bottom right: proof tree, edited specification and program (program view), verification condition (logic view). (Color figure online)

of Fig. 1 is called max; its definition starts with the let keyword. This program computes the maximum of two integers; it consists of a postcondition (that we split into two ensures clauses for readability purposes) and a single instruction. In the postcondition, the result keyword represents the value returned by the program. The window on the left is the proof tree built by Why3; its leaves are the verification conditions (hence the VC keyword) generated by Why3. The node with a Swiss knife shows the strategy used by Why3 to transform a verification condition. In Fig. 1, we have used the default splitting strategy (called split_vc in the proof tree). The second window on the right is called Task; it shows the logical formula associated to a given node in the proof tree (its corresponding specification is highlighted in the program window). For example, in Fig. 1, we have split the postcondition of the program into two subgoals and we have proved with strategy 0 the first subgoal (hence the green bullet in the proof tree). Of course, this is only to illustrate the Why3 graphical user interface. We could have directly (i.e., without splitting) used the strategy 0 to prove that this program satisfies its specification.

3   Learning to Write Precise Specifications

Because there are several possible origins of errors when we specify a program and write its instructions, a main difficulty when teaching deductive verification to beginners is to give them guidelines to understand where these errors come from. Thus, the first part of the course aims at understanding what is a precise specification (i.e., a specification that is precise enough so that the associated instructions can be proved correct, if they indeed satisfy it). Moreover, even if they followed a course on logic before our course, our students have a low background on logic, compared to their background on programming. This section explains how we first ask students to write specifications and to test them. Then, it shows how the testing of specifications scales to larger specifications. Last, it details how we teach students to write precise loop invariants.

3.1 Testing of Specifications

We first learn to specify and write simple programs manipulating integers. The two main goals are to learn the WhyML syntax, and to test the specifications before writing instructions. A test case is a program with an assertion checking that the expected value of the program is satisfied by its specification. Testing a specification is very useful to detect imprecise specifications. Figure 2 shows an example of an imprecise specification and a corresponding test case. The keyword

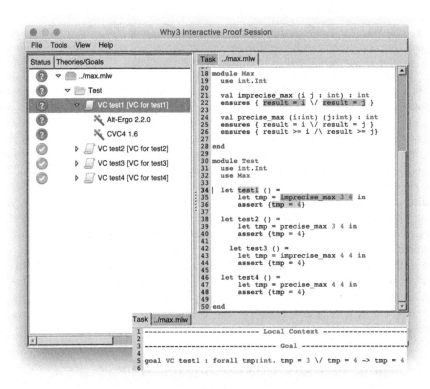

Fig. 2. An example of an imprecise specification (imprecise_max program) and a precise specification (precise_max program).

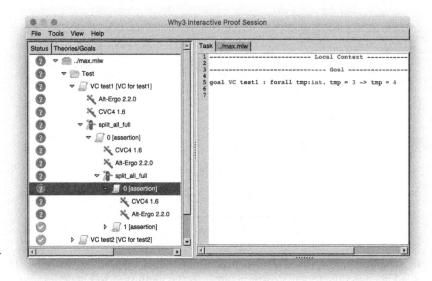

Fig. 3. A logical formula revealing an imprecise specification.

val indicates that the program is only specified (contrary to the keyword let seen in Fig. 1); its instructions will be defined later on. This file consists of two modules: one for the program and another one for the test cases.

The test module consists of the test cases, where each test case is defined in a separate test program. The instructions of a test program are defined (hence the let keyword) but not specified: they call the program to test, and trigger an assertion stating that the program returns the expected value. In WhyML, there is a clear distinction between instructions and logical formulae, which is useful for pedagogical purposes and for using ghost code in a program (see Sect. 4.2). For this reason, only logical formulae can be written in an assertion, which excludes the call to the program to test. We thus use a local definition called tmp to store the result of the tested program.

When we use the basic strategy 0 to prove the test module, the first test case called test1 is not proved. We then switch to the logic window (called Task in Why3, and shown at the bottom of Fig. 2) and look at the corresponding formula, that indeed can not be proved. The formula is the verification condition of test1 : the formula written in the assertion, under the hypothesis that the called program satisfies its specification. When we look at this formula, we understand that $tmp = 4$ can not be proved under the only hypothesis $tmp = 3 \lor tmp = 4$. We must strengthen the specification of imprecise_max into the specification of precise_max .

If we do not yet realize that the formula can not be proved, we can still use the strategy 1 to split as much as possible the formula; here, it is split into two

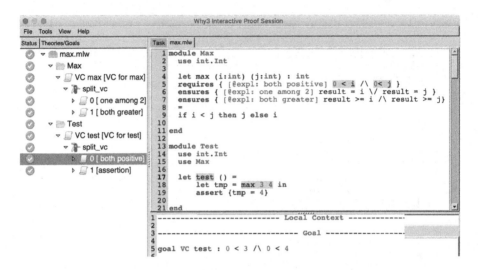

Fig. 4. An example of a proof tree with nodes identified by user explanations. The Task window shows the formula corresponding to the [both positive] node.

subformulae (i.e., a split for each disjunctive case of the hypothesis, as shown by the two Swiss knifes in the proof tree), and try to prove each subformula. As shown in Fig. 3, one subformula remains unproved, and looking again at the logic window will show us a simpler formula (i.e., $tmp = 3 \rightarrow tmp = 4$), that obviously can not be proved.

The take-home message of this example is that the caller p1 of a program p2 only sees the specification of p2 (and not its instructions). So this specification must be precise enough so that p1 can be proved correct. In our test program, the specification of p1 is omitted (meaning that its pre and postconditions are the logical constant true), but the take-home message still holds whatever the specification of p1 is. Moreover, the take-home message generalizes to loop invariants. Indeed, when we teach loop invariants, we face a similar situation, since outside a loop, only the loop invariant is seen. This justifies that the loop invariant needs to be precise as well.

3.2   Testing Larger Specifications

Another way of helping students to understand where an error in the specification comes from is to decompose a specification into a conjunction of subformulae and to annotate each subformula with a short meaningful explanation (e.g., both positive) that will be shown in the proof tree. Figure 4 shows a simple proof tree that is proved correct, once the user has split both verification conditions. His explanations, added in each subformula of its specification annotate the proof tree, so that it becomes easier to connect a particular verification condition with a corresponding instruction in the program. The program max of Fig. 4 computes the same result as the program max of Fig. 1, but with the precondition that its two parameters are strictly positive integers. As shown in the proof tree (where the precondition is called [both positive]), this precondition is verified by the caller of max.

WhyML is a polymorphic language, meaning that the specification language and the programming language use polymorphic types. When Why3 generates the verification conditions of a program, it transforms its polymorphic expressions into monomorphic ones, so that they can be handled by the monomorphic logics of the automated provers that do not support polymorphism. Many data structures defined in the Why3 library (e.g., arrays, lists and trees) use polymorphic types. The first polymorphic data structures we study are arrays of type (array 'a), where 'a denotes a polymorphic type.

Another well-known example of imprecise specification is the specification of a sorting program, and it can be revealed by testing. Specifying only that the array elements are sorted is imprecise, which can be understood by testing the specification. In a precise specification, the property that the sorted array is a permutation of the initial array (i.e., the elements of the sorted array are exactly the elements of the initial array) is required as well. Both properties can be defined by a polymorphic predicate is_sorted (a: array 'a) of the array library that we reuse in several sorting programs.

Students are already familiar with testing, and testing their specifications gives them confidence in their formulae. A typical example of error is the confusion between an implication and a logical and in quantified formulae. Such formulae are extensively used to specify array manipulating programs. Given an array a of integers, an example formula states that all the array cells belong to the interval $[10; 100]$. It can be written as follows, where the predicate (valid_index a i) states that the index i is within the array bounds: $\forall i : int.\ valid\_index\ a\ i \rightarrow 10 \leq a[i] \leq 100$. Indeed, in the specification language of Why3, arrays are modeled using infinite maps from integers to array elements. Defining a property of an array often requires to first select the valid indices of the array.

Testing specification is very useful for students when they fix their first programs. Interestingly, students manage to test their programs on representative test cases (i.e., an array of 20 elements). In previous experiments conducted two years ago, testing the specification of a sorted array was generating formulae that were out of reach of automated provers, for the reason that these formulae were differing too much from the corresponding specification. Once a specification is tested successfully, the next step is to write the instructions of the program.

3.3   Loop Invariants

Without guidance, students tend to write imprecise loop invariants. A false common belief is that the shorter the invariant is, the easier it is proved. The take-home message of Sect. 3.1 is recalled when students write their first loops, and generalized to loop invariants: a loop invariant specifies a loop, and as a specification, it must be precise enough.

First, students write simple loop invariants with loops manipulating only integers. We start with for loops, where the validity of the loop index does not need to be added in the invariant (as it is implicitly handled by Why3), and a loop variant is not needed. The goal is to understand the main parts of a verification condition of a program with a loop. Figure 5 shows an example of a proof tree where we split the verification condition of the loop to see its seven subformulae. The program to prove correct is called max_array; it uses a reference (i.e., a mutable variable) to an integer called m. The syntax for references follows the OCaml syntax, where !m reads the content of reference m and := is used to update this content. max_array computes the maximum element in an array of positive integers and it is adapted from an example of the Why3 gallery of verified programs [16]. Why3 is able to prove directly this program (i.e., without splitting).

Five of the seven subformulae correspond to the three properties that a loop invariant must satisfy: the loop invariant holds at the loop entry (property numbered 0 in the proof tree), it is preserved after any loop iteration (numbers 3 and 4), and it suffices to prove the corresponding postcondition of the program (numbers 5 and 6). These last two properties appear twice in the proof tree because they must hold for any execution of the program (i.e., whether the loop condition initially holds or not), and whether the then branch of the if statement is taken or not. The subformulae numbered 1 and 2 in the proof tree are preconditions of programs called from max_array. Indeed, the cell a[i] is read twice in the program, and this read operation requires the index i to be a valid index.

The task window of Fig. 5 shows the subformula corresponding to the property that the invariant initially (i.e., when the value of !m and i is 0) holds, under the hypotheses called H1 and H2 that the precondition (that we split in two parts in the specification of max_array, hence the two hypotheses) holds. In this simple loop, the loop invariant mimics the postcondition: it is the same property, but related to only some first elements of the array. Thus, we generalized the postcondition into a predicate parameterized by low and high bounds of array indices. We call this predicate is_max and use it in the postcondition and in the loop invariant.

The take-home message of this example is that a loop invariant must be precise enough, and it often mimics the post-condition. Next, students write simple while loops requiring a loop variant to ensure that the loop terminates, and then nested loops. For some programs consisting of a single loop over an array, we ask students to first write the program specification and the loop invariant. Asking them to think of the loop invariant before writing the instructions of the loop makes them write better code (e.g., a single loop instead of nested loops). A well-known example of such a program is the Dutch national flag, which sorts an array of three possible values blue, white and red.

At the end of the course, students write array sorting programs (in increasing order) consisting of two nested loops. We start with a simple array sort (i.e., selection sort) such that the invariants of the two nested loops differ, and the

Fig. 5. An example of loop invariant split into subformulae by Why3.

inner loop (that computes the index of the smallest element in a subarray) can be first proved as an independent program. The invariant of the outer loop is trickier to write, as it is the first example of a loop invariant that partially mimics the postcondition and requires another property that is not expressed in the postcondition. This property states that any array element in the unsorted subarray (i.e., the right part of the array) is greater than any array element in the sorted subarray (i.e., the left part of the array). Some students use the splitting strategy and look at the subformulae that are not proved to understand that a property is missing in the invariant, as introduced in Sect. 3.1. Other students ask for help and we recall them to use the splitting strategy, and ask them to draw on paper the partially sorted array (i.e., a "graphical" loop invariant).

For the more advanced array sorting programs, students need to think in order to discover the inner loop invariant. For the insertion sort, we first give them code of the program that facilitates the writing of the invariant. Then, we ask them to update this code into more efficient code (with less array writes) and to adapt their loop invariant accordingly.

4 Towards More Complex Specifications

When students are familiar with Why3 enough to specify and write simple programs with loop invariants, we explain to them how to build more complex programs. We introduce two features that we describe in this section, type invariants and ghost code.

4.1   Type Invariants

When we study programs involving arrays and matrices, we introduce type invariants. A type invariant follows a type definition and is declared with the keyword invariant ; it is a convenient way of enforcing properties of values of the type. A type invariant is written once for all, thus avoiding to repeat its properties in the specifications. In Why3, type invariants must be ensured at each entry and exit of a program. Moreover, they can only be defined on record types, which explains why we do not use use them at the beginning of our course. This is not a strong limitation, as type invariants become useful when we write more complex programs, with larger specifications involving multiple uses of a same predicate.

Our first example of type invariant is introduced when we study arrays. In the Why3 standard library, a type invariant is used in the definition of arrays to express (among others) that the size of the array is positive. Several of our examples are programs manipulating matrices. A representative example is a maze specified by a matrix of cells, where the cell type consists of the three values denoting a free cell, an exit and a wall of the maze. A type invariant is defined to express properties of the maze, such as there is a single exit. Furthermore, writing programs manipulating matrices gives students another opportunity to write loops and their invariants.

4.2   Ghost Code

Another representative example of a type invariant is a ghost variable to simplify a specification and facilitate a proof. A ghost variable and more generally ghost code do not interfere with the instructions of the program. Our favorite example is the ring buffer data structure. It is available in the gallery of verified programs of Why3 [16]. This program is the third exercise of a challenge given at a verification competition in 2012 [9]; it is detailed in lecture notes written for a Why3 tutorial [8].

The ring buffer is a circular array, starting at a given index in the array and storing N elements, where N is less than the size of the array. In a ring buffer, elements may wrap around the array bounds, as shown in Fig. 6. A new element y will be added after the last added element x, if the array is not full. In this case, if x is stored in the last array cell, then y will be stored in the first array cell. In a similar way, there are two cases to consider when removing an element from a ring buffer. This circularity of the ring buffer makes the array not adapted to specify the ring buffer: there are two cases to consider, depending on whether elements of the ring buffer wrap around the array bounds or not.

The solution to simplify the specification is to use a ghost variable of type sequence that abstracts the ring buffer. There is a one-to-one correspondence between this ghost variable and the array used to implement the ring buffer (see the example in Fig. 6). Sequences are defined in the Why3 standard library. A sequence has no bound, contrary to an array. In a sequence, elements can be added at the end, and removed at its beginning, like in a ring buffer.

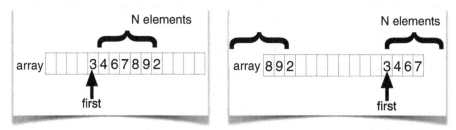

ghost sequence [3; 4; 6; 7; 8; 9; 2]

Fig. 6. An example of two ring buffers consisting of the same elements, implemented in an array, and specified by a same ghost sequence.

The sequence represents in a uniform way the two cases we need to distinguish when using an array in the specification. The correspondence between the sequence and the array is defined once for all in the type invariant of the ring buffer, and the specifications only use the ghost sequence.

This technique is similar to data refinement (used for instance in the B method [1]), and this type invariant can be seen as a gluing invariant between an abstract data (the sequence) and its refinement (the array). Moreover, another advantage of the ghost sequence is that it facilitates the proof of correctness of the operations we define on the ring buffer data structure. To convince students of this fact, we ask them to use different types for the ghost sequence. For example, when the ghost sequence is defined as a list, we observe that the automated solvers need more time to the prove the verification condition of the ring buffer operations.

Last, in the ring buffer program, some instructions need to be added to update the ghost sequence according to the specification. These instructions are ghost code, that must not interfere with regular code (hence the name ghost). When this constraint is not respected, Why3 emits an error message.

5 Proving Recursive Programs

The last part of the course is devoted to recursive programs. We design various algebraic data structures and write associated recursive programs. The match expression with syntax borrowed from OCaml for pattern matching (i.e., filtering the different cases of a sum type) is new for the students. We start with simple recursive programs manipulating integers and polymorphic lists, and then switch to programs manipulating binary trees. When we write a recursive program, we need to define a variant that ensures the termination of the program. A variant is a value that strictly decreases at each recursive call. Well-founded orders are defined in the Why3 standard library for basic types (i.e., integers, lists and trees). So we only need to write in the variant clause the name of the

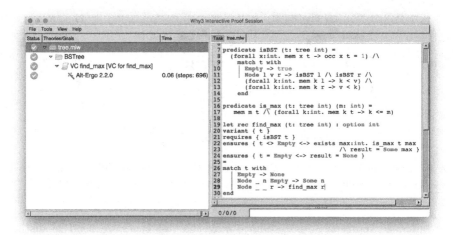

Fig. 7. A recursive program: finding the maximum in a binary search tree.

tree (or of the list or integer) that decreases with respect to its corresponding order. As for loops, the variant generates a verification condition that is easily proved.

The binary tree polymorphic data type is defined in the Why3 library as a type with two constructors called Empty and Node. Figure 7 shows an example of a recursive program called find_max that returns if it exists the maximum element in a binary search tree of integers. The recursive predicate isBST specifies that a tree of integers is a binary search tree; it uses the predicate mem (for member) of the tree library. In this small example, we do not use a type invariant, as we only use isBST once in the precondition. When the tree is empty, the program does not return an integer, hence the value None of the option type to represent this situation. When the tree is not empty, the program returns a (Some v) value, where v is the maximum element in the tree. Recursive calls are done on subtrees of the initial tree t, which ensures the termination of the program, as indicated by the (variant t) clause. The program is proved by Why3 using the basic strategy 0.

Among other examples of recursive programs, we study the well-known example of towers of Hanoi (that is detailed in the Why3 gallery). Next, in other programs, we ask students to write two versions of a program, an imperative one and a recursive one. The students realize that the specification is the same, but the instructions differ and so do the verification conditions. The last kind of recursive programs we study is an imperative program specified by a recursive specification. More precisely, we axiomatize a type or a computation (i.e., we define it in the specification language only by its properties, qualified as axioms as they are not proved by external provers).

6   Conclusion

We presented the Why3 platform and its use to teach a course on deductive verification to undergraduate students. During this course, students managed to specify, program and prove correct non-trivial algorithms such as the selection sort of an array, and the insertion in a binary search tree. Students learned to design data structures and to specify, program and prove correct programs manipulating them. Some of these data structures (ghost variables) facilitate the writing of specifications; they are abstract representations of program variables.

Using Why3 was extremely useful. We were unexpectedly surprised to face many situations where students wrote a meaningful specification together with meaningful instructions but both did not match (e.g., because the specification used $<$ to compare elements but the instructions used $\leq$, or because a variable initialized to 0 in the code was supposed to be initialized to 1 in the specification). Why3 is very useful to reveal these situations and to help students to correct their program so that it satisfies its specification. We only used the automatic strategies 0, 1, 2 and S recently introduced in Why3. They are powerful enough to handle all our programs, which did not require students to change themselves for instance the time-outs of the external provers.

We teach students to first test their specifications using assertions, and they managed to test all of them. Because of the discrepancy between the formulae written in the assertion and the generated verification condition, this would not have been possible two years ago. Another very positive outcome of the improvement and maturity of Why3 and automated provers is that students never needed to add assertions in their programs, so that they could be proved correct. We only use assertions to test a specification, but never to prove a program correct (even if its postcondition is existentially quantified). Adding assertions in a program is a common technique used to help automated provers; several programs of the Why3 gallery (including tricky ones) use it. However, we chose to avoid this technique for beginner students. In the past, we observed that students tend to add as assertions most of the formulae that can not be proved, which most of the time does not solve their problems.

Last, using automated provers is a less and less fragile technology. When students wrote programs manipulating binary search trees, we observed that some programs were more difficult to prove correct than others. For example, if we remove the first pair of parentheses of the predicate isBST defined in Fig. 7, none of the four automated provers we used is able to prove the program find_max. We only faced this situation with programs manipulating binary search trees, and we suspect that the problem stems from the fact that reasoning by induction is difficult for automated provers. To avoid similar situations (that were much more frequent in our past experiences), along the course, we teach students hints to avoid bad practices when writing their specifications. An example of a bad student practice is to always write a quantifier at the beginning of a formula, even if it does not quantify over some parts of the formula.

Acknowledgments. The author would like to thank Delphine Demange and Alan Schmitt for their active participation and their help to prepare this course. The author is grateful to the Why3 development team for its helpful answers, and to Léon Gondelman who convinced her to write this paper.

References

1. Abrial, J.: The B-Book - Assigning Programs to Meanings. Cambridge University Press, Cambridge (1996)
2. Abrial, J., Butler, M.J., Hallerstede, S., Hoang, T.S., Mehta, F., Voisin, L.: Rodin: an open toolset for modelling and reasoning in Event-B. STTT Int. J. Softw. Tools Technol. Transf. **12**(6), 447–466 (2010)
3. Ahrendt, W., et al.: The KeY platform for verification and analysis of Java programs. In: Giannakopoulou, D., Kroening, D. (eds.) VSTTE 2014. LNCS, vol. 8471, pp. 55–71. Springer, Cham (2014). https://doi.org/10.1007/978-3-319-12154-3_4
4. Baudin, P., et al.: ACSL 1.13 documentation. Technical report, CEA (2018)
5. Cormick, J.M., Chapin, P.: Building High Integrity Applications with Spark. Cambridge University Press, Cambridge (2015)
6. Cuoq, P., Kirchner, F., Kosmatov, N., Prevosto, V., Signoles, J., Yakobowski, B.: Frama-C - a software analysis perspective. In: Eleftherakis, G., Hinchey, M., Holcombe, M. (eds.) SEFM 2012. LNCS, vol. 7504, pp. 233–247. Springer, Heidelberg (2012). https://doi.org/10.1007/978-3-642-33826-7_16
7. Filliâtre, J.-C., Paskevich, A.: Why3 — where programs meet provers. In: Felleisen, M., Gardner, P. (eds.) ESOP 2013. LNCS, vol. 7792, pp. 125–128. Springer, Heidelberg (2013). https://doi.org/10.1007/978-3-642-37036-6_8
8. Filliâtre, J.-C.: Deductive program verification with Why3. Digicosme Spring School (2013). http://why3.lri.fr/digicosme-spring-school-2013/
9. Filliâtre, J.-C., Paskevich, A., Stump, A.: VSTTE software verification competition (2012). https://sites.google.com/site/vstte2012/compet
10. Jacobs, B., Smans, J., Philippaerts, P., Vogels, F., Penninckx, W., Piessens, F.: VeriFast: a powerful, sound, predictable, fast verifier for C and Java. In: Bobaru, M., Havelund, K., Holzmann, G.J., Joshi, R. (eds.) NFM 2011. LNCS, vol. 6617, pp. 41–55. Springer, Heidelberg (2011). https://doi.org/10.1007/978-3-642-20398-5_4
11. Leavens, G.T., Baker, A.L., Ruby, C.: JML: a notation for detailed design. In: Kilov, H., Rumpe, B., Simmonds, I. (eds.) Behavioral Specifications of Businesses and Systems, pp. 175–188. Springer, Boston (1999). https://doi.org/10.1007/978-1-4615-5229-1_12
12. Leino, K.R.M.: Dafny: an automatic program verifier for functional correctness. In: Clarke, E.M., Voronkov, A. (eds.) LPAR 2010. LNCS (LNAI), vol. 6355, pp. 348–370. Springer, Heidelberg (2010). https://doi.org/10.1007/978-3-642-17511-4_20
13. Meyer, B.: Applying "design by contract". IEEE Comput. **25**(10), 40–51 (1992)
14. Müller, P., Schwerhoff, M., Summers, A.J.: Viper: a verification infrastructure for permission-based reasoning. In: Jobstmann, B., Leino, K.R.M. (eds.) VMCAI 2016. LNCS, vol. 9583, pp. 41–62. Springer, Heidelberg (2016). https://doi.org/10.1007/978-3-662-49122-5_2
15. Wampler, D., Payne, A.: Programming Scala, 2nd edn. O'Reilly Media, Sebastopol (2014)
16. Gallery of formally verified programs. http://toccata.lri.fr/gallery/why3.en.html

Teaching Program Development

Teaching Formal Methods to Future Engineers

Catherine Dubois[1], Virgile Prevosto[2], and Guillaume Burel[1](✉)

[1] ENSIIE, Samovar, CNRS, Évry, France
{catherine.dubois,guillaume.burel}@ensiie.fr
[2] Institut LIST, CEA, Université Paris-Saclay, Palaiseau, France
virgile.prevosto@cea.fr

Abstract. Formal methods provide systematic and rigorous techniques for software development. We are convinced that they must be taught in Software Engineering curricula. In this paper, we present a set of formal methods courses included in a Software Engineering & Security track of ENSIIE, École Nationale Supérieure d'Informatique pour l'Industrie et l'Entreprise, a French engineering school delivering the «Ingénieur de l'ENSIIE» degree (master level). These techniques have been taught over the last fifteen years in our education programs in different formats. One of the difficulty we encounter is that students consider these kinds of techniques difficult and requiring much work and thus are inclined to choose other courses when they can. Furthermore, students are strongly focused on the direct applicability of the knowledge they are taught, and they are not all going to pursue a professional career in the development of critical systems. Our experience shows that students can gain confidence in formal methods when they understand that, through a rigorous mathematical approach to system specification, they acquire knowledge, skills and abilities that will be useful in their professional future as Computer Scientists/Engineers.

1 Introduction

Formal methods provide systematic and rigorous techniques for reliable software development. Many industries developing critical systems have already adopted formal methods with significant successes (see e.g. [14] for railway experience). Knowing these techniques and methods helps enhancing the quality of software, even in contexts were full-fledged formal verification is not employed. We are thus convinced that formal methods must to be taught in Software Engineering curricula whatever the professional orientation of the future engineers. In this paper, we present a set of formal methods courses included in a Software Engineering & Security track at ENSIIE, École Nationale Supérieure d'Informatique pour l'Industrie et l'Entreprise, a French engineering school delivering the «Ingénieur de l'ENSIIE» degree (master level). These techniques have been taught over the last fifteen years in our education programs in different formats, especially

B. Dongol et al. (Eds.): FMTea 2019, LNCS 11758, pp. 69–80, 2019.
https://doi.org/10.1007/978-3-030-32441-4_5

regarding hourly volumes and elective/compulsory nature. These formal methods courses also reflect a long tradition of research in formal methods at ENSIIE.

The paper is organised as follows. Section 2 presents ENSIIE, its curriculum and specialised tracks. Section 3 quickly introduces the Software Engineering & Security track, emphasizing the courses where formal methods play an important role. Each of these courses is then detailed in a dedicated section (Sects. 4, 5 and 6). We then conclude in Sect. 7.

2   ENSIIE

École Nationale Supérieure d'Informatique pour l'Industrie et l'Entreprise (ENSIIE, https://www.ensiie.fr) is one of the oldest French institutions offering a degree of Engineer (master level) in computer science. Since its creation in 1968, almost 3,000 engineers have graduated from this institution.

Like for the majority of engineering schools in France, most students are admitted at ENSIIE through a selective entrance examination that requires at least two years of preparation with an intensive program in Mathematics and Physics (*Classes Préparatoires aux Grandes Écoles* in French, a very selective curriculum for the first two years in college). ENSIIE hosts about 500 students (around 150 new students per year for a 3-years curriculum).

Students follow a threefold curriculum[1]:

- Information Technology (40%): software engineering, systems and networks, security, artificial intelligence, virtual reality, games and video gaming, robotics, high performance computation;
- Applied Mathematics (30%): operational research, optimisation, data science, machine learning, financial mathematics;
- Business organisation (30%): economy, finance, management, business organisation, entrepreneurship.

A considerable amount of time (11 months during the whole studies) is spent working in companies or research laboratories, corresponding to 3 internships distributed during the study period.

ENSIIE curriculum is organised in 6 semesters or 3 years. Semesters 1 and 2 (first year) form the common core of training with courses in the three main areas previously cited - computer science and engineering, applied mathematics and management - and humanities. This first year has a bachelor level or L3 level according to French educational system. Semesters 3, 4 and 5 mainly correspond to elective technical courses (they are completed by management and humanities). Students can freely choose their courses but specialised tracks are proposed. Because of quotas imposed in some courses, the choices of a student are accepted according to their academic results and personal professional motivations. Semester 6 is dedicated to a 6 months internship leading to a dissertation and a defence evaluated by a jury. These four last semesters end up

---

[1] Course catalogue can be found at https://www.ensiie.fr/wp-content/uploads/2018/05/ensiie_course_catalogue.pdf.

with a master level. During ENSIIE third year (Semesters 5 and 6), students can be enrolled in a research oriented Master (2nd year) in applied mathematics or computer science by attending selected courses from the engineer and master programs. In that case they may have a double degree.

Four specialised tracks are offered: Applied Mathematics (financial analysis, statistics, data science, operational research), Software Engineering & Security (SE & S) (software architecture, systems, formal methods, security), Numerical Interactions (virtual & augmented reality, artificial intelligence), High Performance Computing/Big Data (HPC architecture and operating system, clusters, compilation, numeric simulation). In each track, there are also compulsory and elective courses. Besides, there is also a *free* track in which students are allowed to choose courses from the four previous tracks, composing a menu *à la carte*.

In the rest of the paper, we focus on the SE & S track. Most of the acronyms for courses titles used in this paper stand for names in French. We decided to keep them for a better matching with the official course catalogue.

3   Software Engineering and Security Track

Common core contains some courses related to basics in computer science[2]: imperative programming (C), database design, operating systems, functional programming (OCaml), logic, Web programming (PHP, Javascript) and networks, object oriented programming (Java, C++). Programming projects developed by a team of several students accompany the previous courses.

At this stage, a first formal highlight is given with logic and functional programming. The former course forms the basis for teaching formal methods while the latter introduces students to types, induction, termination and correctness.

Let us focus now on Semesters 3 and 4 (Master 1 level) in SE & S track. S3 courses are mainly compulsory: Agile Project Management, Advanced Functional Programming (IPF), Formal languages, Software Validation and Verification (VVL), Assembly Language and Compilation, Software Engineering (IGL). Students can choose between Operational Research and a course about Security and Protocols. The course entitled IGL introduces students to the principles of Software Engineering and trains them in modelling with UML. It also provides some knowledge about model-driven engineering and quality collaborative project management. Semester S4 is more flexible in the sense that students have some choice, e.g. they can choose between a course about formal methods (MFDLS) and a course about semi numerical algorithms. Then, they have the choice between a course about models of computations (CAL) and a course about the design of privacy-by-design applications. Until 2017, they could also take a class about concurrency and verification by model checking (PCV). However, for structural reasons, this course has not been taught during the last years, but it will be proposed again in 2019–2020, in Semester 4, with a similar content.

Semester 5 (Master 2 level) proposes a large choice of courses. We focus on PROG1 and PROG2 that, among others, belong to the SE & S track.

---

[2] In parenthesis appear the languages used to illustrate the different concepts.

The former focuses on formal proof and formal semantics and the latter on abstract interpretation and deductive verification.

We consider the following set of courses, VVL, MFDLS, CAL, PCV, PROG1, PROG2 as *the formal methods track* or, shorter, *the formal track*. All of them are 42 h long (including lectures, tutorials, lab sessions and exams), except VVL which is only 21 h long. As the number of students is quite low, lectures and tutorials are usually mixed.

VVL introduces students to testing (both black- and white-box testing), and proof of programs (Hoare Logic). Besides lectures and tutorial classes, lab sessions are organized where students use Junit [5], PathCrawler [13] and the Frama-C [12] platform (in particular its deductive verification WP plugin).

CAL, as its name suggests, focuses on calculability and presents several equivalent philosophies and models for computation: Turing machines, partial recursive functions, lambda-calculi. At this point, notions of complexity can be introduced. Eventually, Gödel's first incompleteness theorem is discussed. In order to make these notions more concrete, lab sessions are organized, for example to implement Turing machines.

PCV is concerned with basic concepts of concurrent programming and verification. With these lectures, students acquire in particular the main techniques to verify dynamic properties of concurrent programs (deadlock freeness and more advanced properties) using a model-checker, here SPIN.

We focus below on the three remaining courses, MFDLS, PROG1 and PROG2[3].

In Table 1, we can find the numbers of students that registered in the different courses we focus on in this paper. As mentioned above, the acronyms are related to the French titles. Thus VVL stands for *Software Validation and Verification*, MFDLS for *Formal Methods for Reliable Systems*, CAL for *Models of Computation*, PCV for *concurrency and model-checking*, PROG1 for *Formal Proof and Semantics* and PROG2 for *Static Analysis*. Finally, IGL stands for *Software Engineering*, and IPF for *Advanced Functional Programming*. We can see that these numbers are quite stable over years.

As said before, IGL, IPF and VVL are compulsory courses for the SE & S track. With a very few exceptions (because of the free track), students registered in MFDLS and CAL have been enrolled in VVL in the previous semester. A large number of students take both MFDLS and CAL (50% in 2018–2019). In Semester 5, most of students taking PROG1 have taken MFDLS or at least VVL. PROG1 and PROG2 are taught to the same students, except a very small number of students taking only PROG1.

4   Formal Methods for Reliable Systems (MFDLS)

The course about software validation and verification (VVL) introduces students to formal proofs of programs when programs are annotated with assertions (pre

---

[3] The authors of this paper are teaching these courses.

Table 1. Numbers of students

Semester	Course title	2018–2019	2017–2018	2016–2017
S3	*IGL*	57	57	63
S3	*IPF*	64	69	67
S3	**VVL**	59	59	67
S4	**MFDLS**	25	29	32
S4	**CAL**	19	34	28
S4	**PCV**			29
S5	**PROG1**	17	20	15
S5	**PROG2**	19	18	15

and post-conditions, loop invariants and variants). It is their first encounter with formal specifications. MFDLS makes them go further on that direction with state-based formal methods like B [1] and Event-B [2] and the correct-by-construction development process. The B method was used until spring 2019 when we decided on switch to Event-B. Focus is put on modelling and refinement. We also introduce some security notions, more precisely the main control access policies and show that security issues may also be formalised and integrated to a functional model.

Why moving from B to Event-B while the B version was a well-oiled machine? A first answer would be that Event-B being the *recent* variant of the B method, it should be preferred for teaching newer generations of engineers. However, it is not an easy task. Indeed, B is devoted to developing software with a very long life cycle and it has demonstrated its capacities on large industrial projects (e.g. Paris Meteor line 14), while Event-B is rather a language for modelling systems [14]. However, they share the same foundations: set theory, predicate language, state-based method and refinement. We believe that Event-B refinement is easier and more natural for beginners in formal methods than B refinement. They can understand quite easily the so-called parachute paradigm [2] even if they have difficulties when it comes to implementing it on examples. Finding good gluing invariants remains a hard task, both in B and Event-B. Furthermore, Event-B, with its rather weak language of actions (no if/while substitutions) helps sending the message that modelling and programming are two very distinct activities.

The 2019 schedule is as follows (just replacing Event-B with B will give the previous schedules). Usually, 2 sequences of 3h30 each are scheduled per week on a period of 7 weeks. The first sequence contains an introduction to formal methods and Event-B as well as a presentation of set theory (sets, relations, functions). The second sequence is devoted to - pen and pencil - exercises from simple models requiring only sets to models with sets and relations as variables. For example, one exercice concerns a small system with users that register, log in and log out, revisited with passwords and then with black-listed users. Then, students have a hands-on sequence with Rodin (http://wiki.event-b.org) and

ProB (https://www3.hhu.de/stups/prob/) where they play with or implement some of the models written previously. Faults may have been introduced by the instructor. Sequence 4 is devoted to a formal approach of the semantics of actions and proof obligations. It is also the occasion to review some concepts from logic such as term, formula, free/bound occurrence of a variable and proof rule. Sequence 5 is a lab session, where students learn how to do simple interactive proofs with Rodin. At that time, around the middle of the course, students are evaluated on their ability to manipulate set theory and write some models. This pen-and-pencil evaluation takes place in Sequence 6 and is an hour long. Refinement is then taught and practised during 2,5 sequences with again some practice with Rodin and ProB. A peer-correction of the previous evaluation (described in more details below) takes place meanwhile. For the rest of the course, focus is put on security and control access policies (DAC, MAC and RBAC) with lectures and tutorials. In particular, we study the RBAC encoding (invariants mainly) done within B by Huynh et al. [10] and the combination of a functional model and a security policy. In one of the last sequences, an industrial partner visits us and gives a talk illustrating some real case studies (usually about transportation systems), that motivate students a lot. This talk often opens not only summer internships, but also (and more often) long internships in Semester 6. During the last sequence (Seq. 12), students have to defend their project whose subject has been given in the middle of the course.

The course is illustrated with many examples, from simple to more complex (e.g. Bridge example is studied with the help of Abrial's slides and some youtube videos) giving them good *patterns* to reuse. We encourage both proof and animation though Rodin and ProB but we insist a lot on the differences with respect to verification and validation.

The project has to be realized by 2-persons teams and usually a list of 3 subjects is proposed to the classroom. Most of the projects have security aspects: secure management of medical records, voting system, DAC, ... They are usually case studies inspired by research papers, e.g. in 2019 a reporting management system inspired from [19] that integrates a control access policy close to RBAC but with state-dependent access rights, or a simplified control air traffic control system inspired from [11]. With the description of the system, a refinement plan is proposed. The project is part of the evaluation for 50% of the final mark. Students pass this course with very few exceptions.

Let us come back to the peer correction of the first evaluation that we have been doing for 2 years now. For the moment we do not use any tool for that purpose, so some manual manipulation of assignment papers are required to ensure anonymity of both the corrector and the author. The main benefit for the students is to understand that there are different acceptable solutions. For the teacher it is more work because a solution sheet must be carefully prepared and a double check is necessary. Furthermore, as we do the peer evaluation during a class, the teacher is very much solicited and has to individually help some correctors.

A recurrent difficulty for some students, both in B and in Event-B, is the real nature of invariants and the link with proofs. They do understand proof obligations corresponding to preservation of the invariants by the events. As we noticed while reviewing projects, an informal requirement for a bike sharing system like «A damaged bicycle can not be borrowed by any user» is usually reflected in the pre-condition of an operation modeling the action of borrowing a bike but it is more rarely part of the invariant. As said before in Sect. 1, most of our students have a good background in mathematics. However we can notice that we spend more time to practice set theory and more precisely relational operators because students have less knowledge about that field for some years. We plan to use a set interpreter and an intensive individual training to make it through. For this course, we do not see too much disparity in students' mathematics background. The difference lies in their ability to abstraction.

5 Mechanized Formal Proof and Semantics (PROG1)

This course is equally divided into 2 modules, Mechanized Formal Proof (MFP) and Semantics of Programming Languages (SPL) running in parallel, with one sequence (3h30) for each one in a week. The 2 modules are independent, however the common mathematical tool is the notion of inference rule for proving but also for specifying semantics. Students pass this course with very few exceptions.

MFP is devoted to interactive proving and also automatic proving at an introductory level. Thus, in this module we first step into the Coq interactive theorem prover (https://coq.inria.fr/), used here as an environment to write functional programs, specifications and proofs. We benefit from the fact that our students have studied functional programming and practised OCaml (at least in their common core for most of them), they are used to functions, recursive functions, inductive data types, pattern matching, types and functions as first class values. Hence, they can move from OCaml to Coq quite easily regarding writing code. The first two sequences are hands on, students are introduced to inductively defined predicates and proofs using tactics, up to proofs by induction. At the end of these two sequences, a Coq project is assigned to the students: usually functions on lists (from simple to more elaborate ones, e.g. a simplified version of count-down, sorting function, queue implementation, set as interval list). Projects are done by pairs and must be submitted at the end of the course with a small report using coqdoc.

Then we come back to logic with a reminder of natural deduction for first order logic and a highlight on intuition/classical settings. A quick presentation (which is just a reminder for most students) of pure lambda-calculus and simply typed lambda-calculus (STLC) is done. We then link both worlds by presenting the Curry-Howard (CH) isomorphism. This isomorphism is illustrated on STLC and minimal natural deduction. A blackboard proof is done, describing a process/algorithm to go from a natural deduction proof to a STLC term and back. It is checked on simple examples inside Coq. Then extensions are studied (pairs/conjunction and sum types/disjunction). We do not go further in the

Barendregt cube [3], but we insist on the idea that when logical features are added, the language is extended too. Presenting all this lasts 3 sequences with lectures and tutorials. The part about CH isomorphism is considered as difficult by students. To make it more concrete, we plan to make them implement the production of the lambda term by enriching the tactical prover provided in Chapter 16 of [8].

In the last sequence, students are introduced to automated theorem provers (that they have already encountered in the proof part of VVL when they used the WP Frama-C plugin). We quickly have a look at the DPLL algorithm and implement, during a short lab session, an SMT solver by combining glucose[4] as a SAT solver and glpsol (a tool from the library GLPK[5]) as a solver for linear arithmetic[6].

An exam is organized at the end of the module and the project is part of the evaluation for 50% of the final mark.

The previous module is complemented by a module (SPL) about semantics of programming languages. Students are taught dynamic operational semantics with small step and big step format. Different programming paradigms are revisited (because they have all been practised in other courses in previous semesters). Sequence 1 starts with a language of arithmetic expressions with variables, illustrating the notions of evaluation and environment. Then, we build on this language to formalize the semantics of a small imperative language leading to the notion of execution. Besides tutorials, practical sessions allow students to implement interpreters for the previous languages in OCaml. Then, we move on to a small functional language (Mini-ML) allowing for the introduction of lambda abstractions, closures, and call-by-name vs. call-by-value. Here again, a lab session is organized to develop an OCaml interpreter for Mini-ML, and we also investigate the notion of higher order abstract syntax. A tutorial is usually organized to study other features such as inheritance (using Featherweight Java following the presentation in [17]) or blocks (where locations are introduced). The module ends with a presentation of the K system (http://k-framework. org/) [18], which is an environment for specifying and animating formal semantics, followed by a practical session about this system, going back to the previous simple imperative language.

An exam is organized at the end of the module. Students have to submit the results of some practical sessions, which will account for 30% of the final mark. The main difficulty that the students encounter is the handling of inductive rules that describe the semantics. Although inductive systems are taught already since the logic course of the common core, the students struggle in linking their intuition of the behaviour of programming languages with the design of inductive rules.

---

[4] https://www.labri.fr/perso/lsimon/glucose/.

[5] https://www.gnu.org/software/glpk/.

[6] The lab session text is at the following url http://web4.ensiie.fr/~guillaume.burel/download/PR_TP.pdf.

6   Static Analysis and Deductive Verification (PROG2)

The course contains 6 sequences (3h30 each), giving a brief overview of static analysis and abstract interpretation. It uses a fairly classical minimal imperative language (assignment, test and while loop) as illustration. In parallel students have to work on a project detailed below. Students pass this course with very few exceptions.

The first sequence recalls notions about operational semantics (which in theory have been seen by the students in their previous courses) and introduces the notion of control-flow graphs, concrete execution traces and collecting semantics. The second sequence defines the main grounding blocks of static analysis: lattices and fixpoints, with examples of forward and backward analyses as well as over- and under- approximations. We then move on to Galois connections and insertions and define abstract execution over the sign domain. Widening is seen in the fourth sequence (together with narrowing) and illustrated over intervals. Finally, we present reduced product by showing how the combination of sign and parity information can give more precise results than each piece seen in isolation. The last sequence is dedicated to the presentation of the Eva plugin [7] of Frama-C [12] and a lab session were students use Eva to prove the absence of runtime errors in small C functions, usually extracted from open-source libraries (see for instance https://gitlab.com/vprevosto/stan/wikis/2018-2019/tp for the exercises given this year).

The most important message we try to convey is that it is possible to obtain correct, mathematically backed results about programs, including ones written in real-world languages (hence the last course). As an aside, we also put forward the importance of having precise definitions of the semantics of the various programming languages elements one is working with.

Generally speaking, students do not have a very strong background in logic, which is particularly seen during the lecture on Galois connections and insertions that is usually felt as particularly difficult to grasp.

The main frame of the course is quite stable since the last few years. A small change in the lectures organization has been made possible by the relatively small numbers of students taking the course. While each sequence is formally divided into a lecture followed by a tutorial session, in practice, giving an exercise as soon as the corresponding notion has been introduced proved very beneficial. A more radical change would be to move from pen and paper exercises to lab sessions where they would have to implement these notions, e.g. in OCaml, Why3 or Coq. Such a change would however imply a huge preparation beforehand, and even if students tend to prefer programming rather than doing more theoretical exercises it is not completely clear whether these activities will help them understanding better the theoretical notions that are presented. Indeed a two hours session is very short for proposing something in Coq, or even Why3. On the other hand, an exercise in OCaml would make them focus on an implementation, leaving out the proofs that it is correct. Furthermore, such exercises might interfere with the projects that are described in the next paragraph.

In parallel to the main course, students are asked to work in pairs on a project, consisting in first reading a research article and summarizing it during a short presentation to the whole class, and second doing some software development related to the article. There are usually two categories of subjects for the projects. For each of them, one or two individual subjects are selected, depending on the number of students enrolled in the course. All in all, at most 2 or 3 groups are working on the same subject. The first category is based on an article about static analysis or abstract interpretation and the associated assignment typically consists in implementing the algorithm described in the paper. After many years, where we asked the implementation to take the form of a Frama-C plug-in (or in one occasion of a new domain for Eva), we chose this year to restrict the task to a simple academic language similar to the one presented in the lectures (https://gitlab.com/vprevosto/stan). While letting the students interact with a real framework can be more formative, the complexity of Frama-C's API was a big hurdle to pass before being confronted to the static analysis itself. The two articles this year were Antoine Miné's *A New Numerical Abstract Domain Based on Difference-Bound Matrices* [15] and David Monniaux' and Laure Gonnord's *Cell Morphing: from Array Programs to Array-free Horn Clauses* [16], the latter being probably a bit too ambitious.

The second category is dedicated to deductive verification, with an article on program proofs and a subject consisting in implementing, specifying and proving a small algorithm. Again, this year we shifted from imposing the use of the WP plug-in of Frama-C (and thus a C implementation) to propose Why3 [9], so that students do not have to fight C's idiosyncrasies in addition to think about the best way to write their function contracts and loop invariants. The two articles were *Ghost for Lists: A Critical Module of Contiki Verified in Frama-C* by Allan Blanchard, Nikolai Kosmatov and Frédéric Loulergue [6], and *Secure Information Flow by Self-Composition* by Gilles Barthe, Pedro D'Argenio and Tamara Resk [4]. For the former, the associated subject was the basic operations of the skip list data structure, while for the latter it consisted in the Kruskal algorithm for computing maximal spanning trees over graphs.

7   Conclusion

We presented in this paper a *formal* track offered to students engaged in a Software Engineering & Security curriculum in an engineering school. This has been happening for more than 15 years with variants. Some of our students who have followed this set of courses have a job where they use formal methods every day but a lot of them do not. We interviewed a few of the latter about benefits they got from this formal track in their professional life while they do not apply formal methods directly[7]. To quote one of them, «I think that all the notions we learn about analysis of a program, its source code, and its behaviour, allow us to better understand what we are developing, to better understand what is happening when we write this or that instruction in our code.». And to quote

---

[7] Answers can be found at http://web4.ensiie.fr/~dubois/interviews_FMTEA19.pdf.

another one «Formal methods gave me rigor in software design». We believe that this formal track gives a solid basis to students who want to continue down the *formal* direction (Phd or job relying on formal methods) because it covers a large panel of techniques for specification and verification. For those who go to more traditional development, this formal track gives them rigor, rigor and rigor. This also gives them, when the time comes, the memory that formal tools exist and can help them in a more reliable development.

Acknowledgment. We would like to thank all the colleagues who participated or participate to that set of formal courses. We cite some of them (in any order): S. Blazy, R. Laleau, J. Signoles, X. Urbain, P. Courtieu, F. Gervais, G. Berthelot, A. Mammar, T. Le Gall, R. Rioboo, C. Mouilleron, D. Watel, J. Falampin, C. Métayer, N. Kushik, A. Djoudi. Finally, we mention and thank late P. Facon who introduced a course at ENSIIE about formal specification with VDM in the late 90s and thus opened a specific route.

References

1. Abrial, J.: The B-Book - Assigning Programs to Meanings. Cambridge University Press, Cambridge (2005)
2. Abrial, J.: Modeling in Event-B - System and Software Engineering. Cambridge University Press, Cambridge (2010)
3. Barendregt, H.: Introduction to generalized type systems. J. Funct. Program. **1**(2), 125–154 (1991)
4. Barthe, G., D'Argenio, P.R., Rezk, T.: Secure information flow by self-composition. Math. Struct. Comput. Sci. **21**(6), 1207–1252 (2011)
5. Bechtold, S., Brannen, S., Link, J., Merdes, M., Philipp, M., Stein, C.: J Unit 5 User Guide. https://junit.org/junit5/docs/current/user-guide/
6. Blanchard, A., Kosmatov, N., Loulergue, F.: Ghosts for lists: a critical module of contiki verified in frama-C. In: Dutle, A., Muñoz, C., Narkawicz, A. (eds.) NFM 2018. LNCS, vol. 10811, pp. 37–53. Springer, Cham (2018). https://doi.org/10.1007/978-3-319-77935-5_3
7. Blazy, S., Bühler, D., Yakobowski, B.: Structuring abstract interpreters through state and value abstractions. In: Bouajjani, A., Monniaux, D. (eds.) VMCAI 2017. LNCS, vol. 10145, pp. 112–130. Springer, Cham (2017). https://doi.org/10.1007/978-3-319-52234-0_7
8. Dubois, C., Ménissier-Morain, V.: Apprentissage de la Programmation avec OCaml. Hermès Sciences, Cachan (2004)
9. Filliâtre, J.-C., Paskevich, A.: Why3 — where programs meet provers. In: Felleisen, M., Gardner, P. (eds.) ESOP 2013. LNCS, vol. 7792, pp. 125–128. Springer, Heidelberg (2013). https://doi.org/10.1007/978-3-642-37036-6_8
10. Huynh, N., Frappier, M., Mammar, A., Laleau, R., Desharnais, J.: A formal validation of the RBAC ANSI 2012 standard using B. Sci. Comput. Program. **131**, 76–93 (2016)
11. Jarrar, A., Balouki, Y.: Formal modeling of a complex adaptive air traffic control system. CASM **6**, 6 (2018)
12. Kirchner, F., Kosmatov, N., Prevosto, V., Signoles, J., Yakobowski, B.: Frama-c: a software analysis perspective. Formal Asp. Comput. **27**(3), 573–609 (2015)

13. Kosmatov, N., Williams, N., Botella, B., Roger, M.: Structural unit testing as a service with pathcrawler-online.com. In: SOSE, pp. 435–440. IEEE Computer Society (2013)
14. Lecomte, T., Déharbe, D., Prun, É., Mottin, E.: Applying a formal method in industry: a 25-Year trajectory. In: Cavalheiro, S., Fiadeiro, J. (eds.) SBMF 2017. LNCS, vol. 10623, pp. 70–87. Springer, Cham (2017). https://doi.org/10.1007/978-3-319-70848-5_6
15. Miné, A.: A new numerical abstract domain based on difference-bound matrices. CoRR, abs/cs/0703073 (2007)
16. Monniaux, D., Gonnord, L.: Cell morphing: from array programs to array-free horn clauses. In: Rival, X. (ed.) SAS 2016. LNCS, vol. 9837, pp. 361–382. Springer, Heidelberg (2016). https://doi.org/10.1007/978-3-662-53413-7_18
17. Pierce, B.C.: Types and Programming Languages. MIT Press, Cambridge (2002)
18. Roşu, G., Şerbănuţă, T.F.: An overview of the K semantic framework. J. Logic Algebraic Program. **79**(6), 397–434 (2010)
19. Vistbakka, I., Troubitsyna, E.: Towards Integrated Modelling of Dynamic Access Control with UML and Event-B. arXiv e-prints, May (2018)

The Computational Relevance of Formal Logic Through Formal Proofs

Ariane A. Almeida[1], Ana Cristina Rocha-Oliveira[1],
Thiago M. Ferreira Ramos[1], Flávio L. C. de Moura[1],
and Mauricio Ayala-Rincón[1,2(✉)]

[1] Departamento de Ciência da Computação,
Universidade de Brasília, Brasília, D.F. 70910-900, Brazil
{flaviomoura,ayala}@unb.br
[2] Departamento de Matemática, Universidade de Brasília,
Brasília, D.F. 70910-900, Brazil

Abstract. The construction of correct software, i.e. a computer program that meets a given specification, is an important goal in Computer Science. Nowadays, not only critical software (the ones used in aircraft, hospitals, banks, etc.) is supposed to provide additional guarantees of its correctness. Nevertheless, this is not an easy task because proofs are often long and full of details. In this sense, a strong background in logical deduction is essential to provide Computer Science (CS) professionals the necessary competencies to understand and provide mathematical proofs of their programs. Logic courses for CS tend to follow old precepts without emphasizing mastering *deduction* itself. In our institution, for several years we have followed a more pragmatical approach, in which the foundational aspects of both natural deduction and deduction *à la* Gentzen are taught and, in parallel, the operational premises of deduction are put into practice in proof assistants. Thus, CS students with a minimum knowledge in programming are challenged on providing correctness certificates for simple algorithms. "Putting their hands in the dough" they acquire a better understanding of the value and importance of deductive technologies in computing. Here we show how this is done relating natural deduction and sequent calculus deduction and using the proof assistant PVS in the simple context of a library of sorting algorithms.

1 Introduction

Logic is essential in CS, and the correct manipulation of tools available in this discipline can be very helpful for a good practice of programming and mathematical certification of computational objects. However, motivating the necessity of a profound knowledge about the available deductive frameworks is sometimes hard, if no practical context is provided (to undergraduate students). For doing

Work supported by FAPDF grant 193001369/2016.
M. Ayala-Rincón—Partially supported by CNPq grant 307672/2017-4.

B. Dongol et al. (Eds.): FMTea 2019, LNCS 11758, pp. 81–96, 2019.
https://doi.org/10.1007/978-3-030-32441-4_6

this, we have contextualized our courses on computational logic through the formal verification of basic properties involving simple but relevant algorithms.

Algorithmic properties are verified using the *Prototype Verification System* (PVS), that is a higher-order proof assistant based on sequent calculus (SC) with a functional specification language that supports dependent types. In this way, students are in contact with a deductive tool, used to certify programs and algorithms, directly related to the theory seen in classroom. Our goal is to motivate logic as a relevant branch of CS through the use of simple formalizations.

To challenge students with scarce knowledge on deduction and induction during a one semester course on *Computational Logic*[1], to build complete formalizations would be frustrating and demotivating. This problem is circumvented providing almost complete formalizations of correctness of simple (sorting) algorithms with some strategic *holes*, proposed as conjectures, that are supposed to be completed by them. In this way, students can grasp the connections between theoretical and practical notions needed for fulfilling the demanded tasks. The choice of the proof assistant at this initial level is irrelevant since, after this first contact, having as focus deduction and induction and not formalization, it would not be so difficult for them to migrate to different proof systems.

Even though we believe the importance of relevant results of mathematical logic should not be neglected, such as Gödel's completeness, incompleteness and undecidability theorems, as well as expressiveness bounds of first-order logic and extensions of first-order logic, the most important target when teaching *computational* logic is to provide students enough foundations so that they can master a good understanding of mathematical deduction and computational abilities to apply it in real computational problems.

This work reports how we motivate students to acquire these understanding and abilities with a supervised use of a proof assistant. The textbook that supports the course covers natural deduction (ND) and SC and their equivalence for propositional and predicate (intuitionistic and classical) logic, Gödel completeness theorem, undecidability of the predicate logic and how SC is implemented in PVS [3]. After introducing the required concepts and relations on ND and SC, we introduce PVS functional specifications of sorting algorithms through some easy examples over naturals. These examples follow the lines of the **sorting** library, developed by the authors, and available as part of the NASA LaRC PVS libraries. The **sorting** library includes formalizations of correction of the *Maxsort, Mergesort, Insertion sort, Quicksort, Bubblesort* and *Heapsort* algorithms over elements in a non-interpreted type with an abstract preorder [2].

In Sect. 2, we illustrate how connections between ND and SC are useful for understanding deduction in practical frameworks. In Sect. 3, we show how teaching connections between deductions in a mathematical and in a proof assistant

---

[1] This is a sixteen week (sixty four hour) course exceeding the contents of *Basic Logic* and whose requirements are courses on *Data Structures* and/or *Discrete Structures* according to the CS ACM curricula recommendation. Students typically attend this course on *Computational Logic* after the third semester and, after attending courses on integral and differential calculus, and algebra.

as PVS is valuable for motivating formal deduction. Section 4 shows how induction and recursion, along with formal deduction, can be taught to CS students as a practical tool to analyze correctness of simple recursive algorithms such as basic sorting algorithms. Section 5 presents related work and concludes.

2   Logical Deduction Frameworks

In order to allow students to deal with different deductive styles and proof assistants, we show how to relate deductive frameworks and its rules to prove its equivalence. Here we work with the presentation of the deduction rules *à la* Gentzen for first-order predicate logic (for short SC) and ND, as given in [22]. Then we relate the rules with a few basic proof commands of PVS and its operational semantics. We expect the reader to be familiar with logical deduction, structural induction and recursion.

As a simple example to introduce the notion of relation between frameworks, assume the facts that $(\sqrt{2}^{\sqrt{2}})^{\sqrt{2}}$ and $\sqrt{2}$ are respectively rational and irrational numbers, in order to provide deductions that *there exist irrationals x and y such that x^y is a rational number.* We provide derivations *à la* Gentzen and in natural deduction below, where R is the unary predicate over numbers that is true for rationals, (LEM) means law of excluded middle, sequents are written as $\Gamma \Rightarrow \Delta$ and the symbol ∇ denotes a derivation both in ND and SC. When necessary, we include assumptions (say Γ) and labelled premises (say $[\varphi_1]^{a_1} \dots [\varphi_n]^{a_n}$) above, as well as the conclusion (say φ) of the derivation below this symbol:

$$\Gamma, [\varphi_1]^{a_1} \dots [\varphi_n]^{a_n}$$
$$\nabla$$
$$\varphi$$

The SC derivations ∇_1 and ∇_2 follow:

$$\cfrac{(Ax)\,\neg R(\sqrt{2}^{\sqrt{2}}) \Rightarrow \neg R(\sqrt{2}^{\sqrt{2}}) \qquad \cfrac{\cfrac{\Rightarrow \neg R(\sqrt{2})}{\neg R(\sqrt{2}^{\sqrt{2}}) \Rightarrow \neg R(\sqrt{2})}\,(Lw) \qquad \cfrac{\Rightarrow R((\sqrt{2}^{\sqrt{2}})^{\sqrt{2}})}{\neg R(\sqrt{2}^{\sqrt{2}}) \Rightarrow R((\sqrt{2}^{\sqrt{2}})^{\sqrt{2}})}\,(Lw)}{\neg R(\sqrt{2}^{\sqrt{2}}) \Rightarrow \neg R(\sqrt{2}) \land R((\sqrt{2}^{\sqrt{2}})^{\sqrt{2}})}\,(R_\land)}{\cfrac{\neg R(\sqrt{2}^{\sqrt{2}}) \Rightarrow \neg R(\sqrt{2}^{\sqrt{2}}) \land \neg R(\sqrt{2}) \land R((\sqrt{2}^{\sqrt{2}})^{\sqrt{2}})}{\neg R(\sqrt{2}^{\sqrt{2}}) \Rightarrow \exists x \exists y (\neg R(x) \land \neg R(y) \land R(x^y))}\,(R_\exists)^2}\,(R_\land)$$

$$\cfrac{\cfrac{\cfrac{\Rightarrow \neg R(\sqrt{2})}{R(\sqrt{2}^{\sqrt{2}}) \Rightarrow \neg R(\sqrt{2})}\,(Lw) \qquad \cfrac{\Rightarrow \neg R(\sqrt{2})}{R(\sqrt{2}^{\sqrt{2}}) \Rightarrow \neg R(\sqrt{2})}\,(Lw)}{R(\sqrt{2}^{\sqrt{2}}) \Rightarrow \neg R(\sqrt{2}) \land \neg R(\sqrt{2})}\,(R_\land) \qquad R(\sqrt{2}^{\sqrt{2}}) \Rightarrow R(\sqrt{2}^{\sqrt{2}})\,(Ax)}{\cfrac{R(\sqrt{2}^{\sqrt{2}}) \Rightarrow \neg R(\sqrt{2}) \land \neg R(\sqrt{2}) \land R(\sqrt{2}^{\sqrt{2}})}{R(\sqrt{2}^{\sqrt{2}}) \Rightarrow \exists x \exists y (\neg R(x) \land \neg R(y) \land R(x^y))}\,(R_\exists)^2}\,(R_\land)$$

And from ∇_1 and ∇_2 above, one has:

$$\cfrac{(LEM)\,\Rightarrow \neg R(\sqrt{2}^{\sqrt{2}}) \lor R(\sqrt{2}^{\sqrt{2}}) \qquad \cfrac{\nabla_1 \qquad \nabla_2}{\neg R(\sqrt{2}^{\sqrt{2}}) \lor R(\sqrt{2}^{\sqrt{2}}) \Rightarrow \exists x \exists y (\neg R(x) \land \neg R(y) \land R(x^y))}\,(L_\lor)}{\Rightarrow \exists x \exists y (\neg R(x) \land \neg R(y) \land R(x^y))}\,(cut)$$

Now, the ND derivations ∇'_1 and ∇'_2 follow:

$$\cfrac{\cfrac{\cfrac{\neg R(\sqrt{2}) \quad R((\sqrt{2}^{\sqrt{2}})^{\sqrt{2}})}{[\neg R(\sqrt{2}^{\sqrt{2}})]^{a_1} \quad \neg R(\sqrt{2}) \wedge R((\sqrt{2}^{\sqrt{2}})^{\sqrt{2}})}\,(I_\wedge)}{\neg R(\sqrt{2}^{\sqrt{2}}) \wedge \neg R(\sqrt{2}) \wedge R((\sqrt{2}^{\sqrt{2}})^{\sqrt{2}})}\,(I_\wedge)}{\exists x \exists y (\neg R(x) \wedge \neg R(y) \wedge R(x^y))}\,(I_\exists)^2$$

$$\cfrac{\cfrac{\cfrac{\neg R(\sqrt{2}) \quad \neg R(\sqrt{2})}{\neg R(\sqrt{2}) \wedge \neg R(\sqrt{2})}\,(I_\wedge) \quad [R(\sqrt{2}^{\sqrt{2}})]^{a_2}}{\neg R(\sqrt{2}) \wedge \neg R(\sqrt{2}) \wedge R(\sqrt{2}^{\sqrt{2}})}\,(I_\wedge)}{\exists x \exists y (\neg R(x) \wedge \neg R(y) \wedge R(x^y))}\,(I_\exists)^2$$

And from these derivations, one obtains:

$$\cfrac{(\text{LEM})\ \neg R(\sqrt{2}^{\sqrt{2}}) \vee R(\sqrt{2}^{\sqrt{2}}) \qquad \nabla'_1 \qquad \nabla'_2}{\exists x \exists y (\neg R(x) \wedge \neg R(y) \wedge R(x^y))}\,(E_\vee, a_1, a_2)$$

It is straightforward that SC right rules and ND introduction rules perform corresponding tasks in both frameworks. For instance, observe the second application of rules $(R_\wedge)$ and $(I_\wedge)$ in the derivations ∇_1 and ∇'_1 above. On the other hand, understanding the relation between SC left rules and ND elimination rules requires a good comprehension of the role of (cut) in SC. For instance, observe the application of $(L_\vee)$ and (cut) and $(E_\vee)$ in previous derivations. The correspondence between ND and SC rules is the basis to verify that restricted to the intuitionistic fragment both derivation mechanism are equivalent. Below we illustrate the necessary translation for the specific case of derivation rules $(E_\vee)$ and (cut) plus $(L_\vee)$. In these derivations, the symbols ∇ and ∇' (as well as their subscripted versions) represent derivations in ND and SC, and a derivation ∇' is inductively assumed from the existence of the corresponding derivation ∇.

$$\cfrac{\begin{array}{ccc} \Gamma & [\phi]^a\ \Gamma & [\varphi]^b\ \Gamma \\ \nabla & \nabla_1 & \nabla_2 \\ \phi \vee \varphi & \xi & \xi \end{array}}{\xi}\,(E_\vee, a, b) \qquad \rightsquigarrow \qquad \cfrac{\cfrac{\nabla'}{\Gamma \Rightarrow \phi \vee \varphi} \qquad \cfrac{\cfrac{\nabla'_1}{\phi, \Gamma \Rightarrow \xi} \quad \cfrac{\nabla'_2}{\varphi, \Gamma \Rightarrow \xi}}{\phi \vee \varphi, \Gamma \Rightarrow \xi}\,(L_\vee)}{\Gamma \Rightarrow \xi}\,(\text{cut})$$

For the equivalence for the whole classical first-order logic, additional problems arise, such as the restriction of sequents in intuitionistic logic to have at most one formula in their succedent. This is solved by applications of c-equivalent transformations steps on sequents, a discipline also used in proof assistants as PVS to provide sequents only with positive formulas.

To prove that $\neg\phi, \Gamma \Rightarrow \Delta$ implies $\Gamma \Rightarrow \Delta, \phi$ one uses non-intuitionistic stability axioms $(\Rightarrow \neg\neg\phi \rightarrow \phi)$ derived from axioms (Ax) and $(L\bot)$ as shown in the left sub-branch below.

$$\cfrac{\cfrac{\cfrac{\cfrac{\phi \Rightarrow \phi, \bot\ (\text{Ax})}{\Rightarrow \phi, \neg\phi}\,(R_\rightarrow) \quad \bot \Rightarrow \phi\ (L\bot)}{\neg\neg\phi \Rightarrow \phi}\,(L_\rightarrow)}{\Rightarrow \neg\neg\phi \rightarrow \phi}\,(R_\rightarrow) \qquad \cfrac{\cfrac{\cfrac{\neg\phi, \Gamma \Rightarrow \Delta}{\neg\phi, \Gamma \Rightarrow \Delta, \phi, \bot}\,(R_W)}{\Gamma \Rightarrow \Delta, \phi, \neg\neg\phi}\,(R_\rightarrow) \quad \phi, \Gamma \Rightarrow \Delta, \phi\ (\text{Ax})}{\neg\neg\phi \rightarrow \phi, \Gamma \Rightarrow \Delta, \phi}\,(L_\rightarrow)}{\Gamma \Rightarrow \Delta, \phi}\,(\text{cut})$$

The other direction is obtained using the sequents $\Gamma \Rightarrow \Delta, \phi$ and $\bot, \Gamma \Rightarrow \Delta$ as premises of an application of rule $(L_\rightarrow)$. The second premise is obtained through application of rule $(L_\perp)$. The proof for the sequents $\phi, \Gamma \Rightarrow \Delta$ and $\Gamma \Rightarrow \Delta, \neg\phi$ does not require stability axioms.

We denote as (c-eq) all four invertible derived rules related with c-equivalence:

$$\frac{\neg\phi, \Gamma \Rightarrow \Delta}{\Gamma \Rightarrow \Delta, \phi} \text{ (c-eq)} \qquad \frac{\Gamma \Rightarrow \Delta, \neg\phi}{\phi, \Gamma \Rightarrow \Delta} \text{ (c-eq)}$$

3   Logical Deduction versus Proof Commands

Understanding how proof assistants use the chosen deductive calculus is necessary in a more practical setting. In PVS for instance, that relation is clarified by highlighting the main differences between its proofs and SC derivations.

The first observation is that derivations in PVS start from the conclusion, thus, the proof trees can be seen as SC derivations upside down. Another point is that propositional steps and applications of properties such as (LEM) and (c-eq) as well as axioms are done automatically by the proof assistant, while in SC derivations this has to be done step by step.

To provide a formal proof about some conjecture with PVS, first one specifies it using the specification language. Then, the proving process starts using the proof language whose proof commands follow a SC style [14,15].

To illustrate this we examine a PVS deduction tree for the example of the previous section, specified as the following conjecture, where R? abbreviates the predicate **rational?**. The root node (see Fig. 1) is labelled by the objective sequent, but for simplicity, all sequents were dropped from this tree:

⊢ EXISTS (x,y) : NOT R?(x) AND NOT R?(y) AND R?(x^y)

As first derivation step, related with SC rule (cut), one must proceed by case analysis using the command (case), providing two branches (see Fig. 1).

The right branch brings as objective the instance of (LEM) used in the case, that is ''R?(^ ($\sqrt{2},\sqrt{2}$)) OR NOT R?(^ ($\sqrt{2},\sqrt{2}$))'' as a succedent, which is proved using rule $(R_\vee)$, applied through the PVS proof command (flatten), generating two succedent formulas, R?(^ ($\sqrt{2},\sqrt{2}$)) and NOT R?(^ ($\sqrt{2},\sqrt{2}$)). The system automatically applies (c-eq) and concludes this branch with a sole command, by moving the second formula to the antecedent without negation, and then (Ax). In the left branch that is the interesting one, PVS provides the formula referent to the (case) as an antecedent. Thus, rule $(L_\vee)$ should be applied to split the proof into two sub-branches, which is done by PVS proof command (split) along with c-equivalence.

The left sub-branch brings a sequent with the formula R?(^ ($\sqrt{2},\sqrt{2}$)) in the antecedent and is related to derivations ∇_2 and ∇_2', while the right sub-branch is related to derivations ∇_1 and ∇_1' and has this formula in the succedent

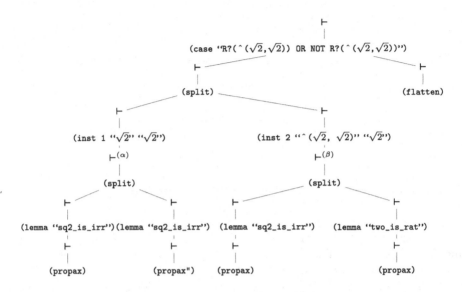

Fig. 1. PVS deduction tree for existence of irrationals with rational potentiation

due to c-equivalence. In both cases, one must deal adequately with the existential quantifiers in the target conjecture EXISTS (x,y) : NOT R?(x) AND NOT R?(y) AND R?(^ (x,y)) with the SC rule ($R_\exists$). Since it is done in a down to up manner, concrete and adequate witnesses of the existence of these quantified variables should be provided, requiring from students a good comprehension of quantifier inference rules and also creativity to find the right instantiations.

For the left sub-branch, applying instantiation (by command (inst ''$\sqrt{2}$" "$\sqrt{2}$'')) gives as result the sequent (label of the node marked with (α))

$$R?(^(\sqrt{2},\sqrt{2})) \vdash \text{NOT } R?(\sqrt{2}) \text{ AND NOT } R?(\sqrt{2}) \text{ AND } R?(^(\sqrt{2},\sqrt{2}))$$

The formula in the succedent splits into three objectives by application of ($R_\wedge$) using the PVS proof command (split). The third of these objectives is trivially discharged by an automatic application of (Ax), since R?(^ ($\sqrt{2},\sqrt{2}$)) is also an antecedent of the sequent. The other two require the knowledge that NOT R?($\sqrt{2}$), obtained from a lemma as (lemma ''sq2_is_irr'').

For the right sub-branch, the instantiation (inst '' ^ ($\sqrt{2},\sqrt{2}$)'' ''$\sqrt{2}$'') generates the sequent (label of the node (β))

$$\vdash \text{NOT } R?(^(\sqrt{2},\sqrt{2})) \text{ AND NOT } R?(\sqrt{2}) \text{ AND } R?(^(^(\sqrt{2},\sqrt{2}), \sqrt{2})), R?(^(\sqrt{2},\sqrt{2}))$$

The proof proceeds as in the left sub-branch, requiring (lemma ''two_is_rat'') to state that R?(^ (^ ($\sqrt{2},\sqrt{2}$) $\sqrt{2}$) to complete its last sub-branch.

In general, when dealing with quantifiers, the proof command related with both rules ($R_\exists$) and ($L_\forall$) is (inst), while for both rules ($R_\forall$) and ($L_\exists$) what is required is application of Skolemization through the proof command (`skolem`). In the last case, PVS substitutes quantified variables by *fresh* ones.

When using proof assistants to deal with algorithmic properties, the syntax of branching instructions of the specification language such as `IF-THEN-ELSE` need to be related to logic by establishing their concrete semantics. For instance, by illustrating `IF-THEN-ELSE` instruction behavior in PVS one has:

$$\frac{a \vdash b}{\vdash a \to b}\ \text{(flatten)} \qquad \frac{\vdash a, c}{\dfrac{\vdash \neg a \to c}{\vdash\ \text{IF}\ a\ \text{THEN}\ b\ \text{ELSE}\ c\ \text{ENDIF}}}\ \begin{array}{l}\text{(flatten)}\\ \text{(split)}\end{array}$$

$$\frac{a, b \vdash}{\dfrac{a \wedge b \vdash}{\text{IF}\ a\ \text{THEN}\ b\ \text{ELSE}\ c\ \text{ENDIF} \vdash}}\ \text{(flatten)} \qquad \frac{c \vdash a}{\neg a \wedge c \vdash}\ \begin{array}{l}\text{(flatten)}\\ \text{(split)}\end{array}$$

Table 1 summarises a few relations between SC and ND rules and PVS proof commands. Marks in the second column indicate that rules (Ax), $(L_\bot)$ and (c-eq) are automatically applied whenever possible along with proof commands.

Table 1. SC and ND rules versus PVS proof commands

	$(Ax)(L_\bot)$ c-equiv	$(LR_W)(LR_C)$	$(L_\wedge)$ $(E_\wedge)$	$(L_\vee)$ $(E_\vee)$	$(L_\to)$ $(E_\to)$	$(L_\forall)$ $(E_\forall)$	$(L_\exists)$ $(E_\exists)$	$(R_\wedge)$ $(I_\wedge)$	$(R_\vee)$ $(I_\vee)$	$(R_\to)$ $(I_\to)$	$(R_\forall)$ $(I_\forall)$	$(R_\exists)$ $(I_\exists)$	(cut)
(hide)		×											
(copy)		×											
(flatten)	✓		×						×	×			
(split)	✓			×	×			×					
(skolem)	✓						×				×		
(inst)	✓					×						×	
(lemma)													×
(case)	✓												×

4 Inductive Proofs Versus Recursive Algorithms

To move to the formalization of correctness of simple algorithms as a mechanism to motivate students in the study of formal logic and deduction, other important concepts, such as recursion and induction are necessary. As example, we propose the formalization of correctness of sorting algorithms, as will be illustrated using Hoare's Quicksort algorithm specified over lists of a non-interpreted type `T` in which a preorder is available. Lists are specified as usual as an inductive data structure where `null` is the constructor for empty lists, and `cons` constructs a new list from a given element of `T` and a list. The operators `cdr` and `car` give the tail and head of a list: `cdr(cons(x,l)) := l` and `car(cons(x,l)) := x`.

```
quick_sort (l : list[T]): RECURSIVE list[T] =
    CASES l OF null: null,
        cons (x, r): append(quick_sort (leq_elements (r,x)),
                            cons(x,quick_sort(g_elements(r, x)))))
    ENDCASES
    MEASURE length(l)
```

Above, the head of the list is chosen as the pivot x; leq_elements(r,x) and g_elements(r,x) build the lists of elements of the tail r that are respectively less than or equal to and greater than x; these lists are then recursively sorted and properly appended, along with the pivot, to provide the sorted list. The termination MEASURE for this function is given as the length of the list. This measure helps one to conclude that quick_sort is well-defined (and terminating) by proving that each recursive call has as argument a list whose length is strictly smaller than the input list. Sometimes PVS can prove well-definedness automatically, but in general, as in this case, the user needs to prove that the measure indeed decreases. The proof for Quicksort uses the following lemmas:

```
leq_elements_size : LEMMA            g_elements_size : LEMMA
FORALL (l : list[T], x:T) :          FORALL (l : list[T], x:T) :
  length(leq_elements(l,x)) <=          length(g_elements(ll,x)) <=
            length(l)                            length(l)
```

Proving these lemmas is left as exercise that requires structural induction, which is usually not easily digested by the students. In fact, their first practical contact with structural induction happens likely in this course when they have to prove, for instance, that well-formed expressions in the syntax of first-order logic satisfy some properties or, when they have to prove by induction in derivations the correctness of ND or of SC. Similarities with natural induction might be used to explain this principle. Structural induction principle of PVS is used through command (induct) followed by the induction variable. In general, with P as the predicate representing the property to be proved, one has the following schema:

$$\Gamma \vdash \forall(l) : P(l), \Delta$$

$$\text{(induct ``l'')}$$

$$\Gamma \vdash P(\text{null}), \Delta \qquad \Gamma \vdash \forall(x, l') : P(l') \to P(\text{cons}(x, l')), \Delta$$

More interesting, a complete or strong induction principle can be applied, where a different measure, say μ, from the one extracted from the inductive data structure is used, which is built as the following schema:

$$\Gamma \vdash \forall(l) : P(l), \Delta$$

$$\text{(measure-induct ``}\mu(l)\text{'' (``l''))}$$

$$\Gamma \vdash \forall(l) :(\forall(x, l') : (\mu(l') < \mu(l) \to P(l'))) \to P(l), \Delta$$

After applying proof commands (skolem) and (flatten), one has the sequent $\Gamma, (\forall(x, l') : (\mu(l') < \mu(l) \to P(l'))) \vdash P(l), \Delta$, which can also be obtained by the PVS command (measure-induct+) with the same arguments as above.

Using structural induction on l to prove lemma leq_elements_size above, one obtains the sub-objectives below, where leq_l abbreviates leq_elements:

```
|- FORALL (x: T): length(leq_l(null, x)) <= length(null)
```

and

```
|- FORALL (x': T, l': list[T]):
   (FORALL (x: T):  length(leq_l(l', x)) <= length(l'))
    IMPLIES  (FORALL (x: T):  length(leq_l(cons(x', l'), x)) <=
                                    length(cons(x', l')))
```

For proving the former, it is enough to apply the command (skolem) and then expand the definition of leq_elements obtaining the trivial sequent ⊢length(null) <= length(null). For proving the latter, after applying the commands (skolem) twice and then (flatten), one obtains the sub-objective:

```
FORALL (x: T):  length(leq_l(l', x)) <= length(l')
 |- FORALL (x: T):  length(leq_l(cons(x', l'), x)) <= length(cons(x', l'))
```

After Skolemization, instantiation and expansions of length and leq_l (i.e., leq_elements), one obtains the following objective:

```
length(leq_l(l', x)) <= length(l')
 |- IF x' <= x THEN 1 + length(leq_l(l', x))
     ELSE length(leq_l(l', x))  ENDIF   <= 1 + length(l')
```

Notice that the IF-THEN-ELSE expression can be *lifted* to the top of the succedent (using proof command (lift-if)), obtaining the equivalent formula

```
IF x' <= x THEN 1 + length(leq_l(l', x))    <= 1 + length(l')
ELSE length(leq_l(l', x))    <= 1 + length(l')  ENDIF
```

According to the discussion on the semantics of IF-THEN-ELSE, it is possible to apply proof commands obtaining two sub-objectives that can be concluded directly applying the command (assert), which will automatically apply the necessary simplifications from the PVS prelude library using decision procedures.

Notice that when performing this kind of exercise, justifying the application of several proof techniques (or proof commands) will demand additional effort, for instance, to let clear to students that, when definitions are expanded (through proof command (expand)), an equational derivation rule is being applied. Such rule essentially replaces the application of an operator by its definition, exactly as it was defined in the specification. Similarly, other equational commands that are required for equational management, such as (replace) which replaces the left or right-hand side of an equation for the other side of the equation in other formulas of the sequent, can be explained. These deduction mechanisms could be easily schematized and explained as the SC rules:

$$\frac{(\Gamma \Rightarrow \Delta)[f(t) \mapsto def_f(t)]}{\Gamma \Rightarrow \Delta} \ (L_=) \qquad \frac{\Gamma[s \mapsto t], s \doteq t \Rightarrow \Delta[s \mapsto t]}{\Gamma, s \doteq t \Rightarrow \Delta} \ (L_=) \qquad \overline{\Rightarrow x = x} \ (R_=)$$

where, $\doteq$ denotes that the orientation of the equational formula is irrelevant, and $[s \mapsto t]$ denotes the *substitution* of some occurrences of s by t. And $f(t)$ denotes an application of the operator f to the argument(s) t and $def_f(t)$ denotes the instantiation of the body of the specification of the operator f with t. In rule

$(L_=)$, when all occurrences of s are replaced by t, the equation $s \doteq t$ in the antecedent of the premise sequent might be dropped.

Regarding *lifting* IF–THEN–ELSE's to the top of the formulas, it can be seen as an invertible SC rule that, according to the semantics of this branching instruction, corresponds to the command below, where P is a predicate on tuples of the type of b (that should be the same type of c) and the type of d. This rule behaves identically in the antecedent of a sequent.

$$\frac{\Gamma \vdash \texttt{P(IF } a \texttt{ THEN } b \texttt{ ELSE } c \texttt{ ENDIF, } d), \ \Delta}{\Gamma \vdash \texttt{IF } a \texttt{ THEN } \texttt{P}(b,d) \texttt{ ELSE } \texttt{P}(b,c) \texttt{ ENDIF}, \Delta} \ (\texttt{lift-if})$$

Translating complete PVS proofs to SC derivations helps to consolidate the understanding of relations between theory and practice. To illustrate this, consider the simple property that list that are permutations are also permutations after adding to them the same element. Permutations? (for short Perm?) is a predicate that appears in the main correctness theorem of Quicksort below.

```
quick_sort_works: THEOREM    FORALL (1 : list[T]):
  is_sorted?(quick_sort(1)) AND Permutations?(quick_sort(1), 1)
```

Perm? is specified over pairs of lists based on the coincidence of occurrences of elements in the lists: $\texttt{Perm?}(l_1, l_2) := \forall x \, \texttt{O}(l_1, x) = \texttt{O}(l_2, x)$, where $\texttt{O}(l, x)$ is specified as the number of occurrences of x in l. Assume the knowledge that $\Rightarrow \forall y \, \texttt{O}(y \cdot l, y) = 1 + \texttt{O}(l, y) \equiv \varphi(l)$ and $\Rightarrow \forall x, y \, \neg x = y \rightarrow \texttt{O}(x \cdot l, y) = \texttt{O}(l, y) \equiv \psi(l)$, where $x \cdot l$ abbreviates $\texttt{cons}(x, l)$. In order to provide deductions to prove that adding the same element x to permutations l_1 and l_2 one obtains lists $x \cdot l_1$ and $x \cdot l_2$ that are also permutations. Below, a SC derivation that follows the steps of the PVS proof is given.

$$\frac{\dfrac{\overset{\nabla_1 \qquad\qquad \nabla_2}{(LEM) \Rightarrow x = y \vee \neg x = y \quad x = y \vee \neg x = y, \texttt{O}(l_1, y) = \texttt{O}(l_2, y) \Rightarrow \texttt{O}(x \cdot l_1, y) = \texttt{O}(x \cdot l_2, y)}}{\dfrac{\texttt{O}(l_1, y) = \texttt{O}(l_2, y) \Rightarrow \texttt{O}(x \cdot l_1, y) = \texttt{O}(x \cdot l_2, y)}{\dfrac{\forall y \, \texttt{O}(l_1, y) = \texttt{O}(l_2, y) \Rightarrow \texttt{O}(x \cdot l_1, y) = \texttt{O}(x \cdot l_2, y)}{\dfrac{\forall y \, \texttt{O}(l_1, y) = \texttt{O}(l_2, y) \Rightarrow \forall y \, \texttt{O}(x \cdot l_1, y) = \texttt{O}(x \cdot l_2, y)}{\dfrac{\texttt{Perm?}(l_1, l_2) \Rightarrow \texttt{Perm?}(x \cdot l_1, x \cdot l_2)}{\dfrac{\Rightarrow \texttt{Perm?}(l_1, l_2) \rightarrow \texttt{Perm?}(x \cdot l_1, x \cdot l_2)}{\Rightarrow \forall l_1, l_2, x \, \texttt{Perm?}(l_1, l_2) \rightarrow \texttt{Perm?}(x \cdot l_1, x \cdot l_2)} \, (R_\forall)^3}}{} \, (R_\rightarrow)}{} \, (L_=)}{} \, (R_\forall)}{} \, (L_\forall)}}{} \begin{matrix}(L_\vee) \\ (cut)\end{matrix}$$

Where ∇_1 is given as:

$$\frac{\Rightarrow \varphi(l_1) \qquad \nabla_1'}{\dfrac{\texttt{O}(l_1, y) = \texttt{O}(l_2, y) \Rightarrow \texttt{O}(y \cdot l_1, y) = \texttt{O}(y \cdot l_2, y)}{x = y, \texttt{O}(l_1, y) = \texttt{O}(l_2, y) \Rightarrow \texttt{O}(x \cdot l_1, y) = \texttt{O}(x \cdot l_2, y)} \, (L_=)} \, (cut)$$

with ∇_1':

$$\frac{\Rightarrow \varphi(l_2) \qquad \dfrac{\dfrac{\dfrac{\Rightarrow 1 + \texttt{O}(l_1, y) = 1 + \texttt{O}(l_1, y)}{\texttt{O}(l_1, y) = \texttt{O}(l_2, y) \Rightarrow 1 + \texttt{O}(l_1, y) = 1 + \texttt{O}(l_2, y)} \, (L_=)}{\texttt{O}(y \cdot l_2, y) = 1 + \texttt{O}(l_2, y), \texttt{O}(l_1, y) = \texttt{O}(l_2, y) \Rightarrow 1 + \texttt{O}(l_1, y) = \texttt{O}(y \cdot l_2, y)} \, (L_=)}{\varphi(l_2), \texttt{O}(l_1, y) = \texttt{O}(l_2, y) \Rightarrow 1 + \texttt{O}(l_1, y) = \texttt{O}(y \cdot l_2, y)} \, (L_\forall) \, R_=}{\dfrac{\texttt{O}(l_1, y) = \texttt{O}(l_2, y) \Rightarrow 1 + \texttt{O}(l_1, y) = \texttt{O}(y \cdot l_2, y)}{\dfrac{\texttt{O}(y \cdot l_1, y) = 1 + \texttt{O}(l_1, y), \texttt{O}(l_1, y) = \texttt{O}(l_2, y) \Rightarrow \texttt{O}(y \cdot l_1, y) = \texttt{O}(y \cdot l_2, y)}{\varphi(l_1), \texttt{O}(l_1, y) = \texttt{O}(l_2, y) \Rightarrow \texttt{O}(y \cdot l_1, y) = \texttt{O}(y \cdot l_2, y)} \, (L_\forall)} \, (L_=)}{} \begin{matrix}(cut)\end{matrix}}{} \, (L_=)$$

And ∇_2 is given as:

$$\Rightarrow \psi(l_1)$$
$$\vdots$$

$$\nabla_2 : \quad \cfrac{\cfrac{\Rightarrow x = y, 0(x \cdot l_1, y) = 0(l_1, y) \qquad \nabla_2'}{0(l_1, y) = 0(l_2, y) \Rightarrow x = y, 0(x \cdot l_1, y) = 0(x \cdot l_2, y)} \text{(cut)}}{\neg x = y, 0(l_1, y) = 0(l_2, y) \Rightarrow 0(x \cdot l_1, y) = 0(x \cdot l_2, y)} \text{(c-eq)}$$

with ∇_2':

$$\Rightarrow \psi(l_2)$$
$$\vdots$$

$$\cfrac{\cfrac{\Rightarrow x = y, 0(x \cdot l_2, y) = 0(l_2, y) \quad \cfrac{0(l_1, y) = 0(l_2, y) \Rightarrow 0(l_1, y) = O(l_2, y) \text{ (Ax)}}{0(x \cdot l_2, y) = 0(l_2, y), 0(l_1, y) = 0(l_2, y) \Rightarrow 0(l_1, y) = 0(x \cdot l_2, y)} \begin{matrix}(L_=)\\(\text{cut})\end{matrix}}{0(l_1, y) = 0(l_2, y) \Rightarrow x = y, 0(l_1, y) = 0(x \cdot l_2, y)}}{0(x \cdot l_1, y) = 0(l_1, y), 0(l_1, y) = 0(l_2, y) \Rightarrow x = y, 0(x \cdot l_1, y) = 0(x \cdot l_2, y)} (L_=)$$

The proof that Quicksort computes a permutation starts with the sequent below (labelling the root node $\vdash_0$ in Fig. 2):

```
|- FORALL (l : list[T]): Perm?(quick_sort(l),l)
```

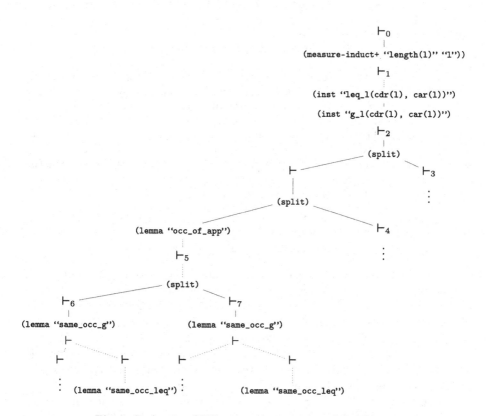

Fig. 2. Deduction PVS tree to permutation of Quicksort

Now, by strong induction applying (measure-induct+) using as measure the length of lists, the sequent below is obtained ($\vdash_1$ in Fig. 2), where the induction hypothesis is the antecedent formula (type annotation is omitted by short):

```
FORALL(y): length(l') < length(l) IMPLIES Perm?(quick_sort(l'),l')
|- Perm?(quick_sort(l),l)
```

After that, proper applications of (expand) as well as (copy) and (inst) are done. In particular, for the last proof command, two adequate instantiations (that is, SC rule ($L_\forall$)) of the induction hypothesis are needed: one with the sublist containing the elements smaller than or equal the pivot and the other with the sublist containing the elements greater than the pivot, giving the sequent below ($\vdash_2$ in Fig. 2), where for brevity q_s, g_l and app respectively abbreviate quick_sort, g_elements and append.

```
length(g_l(cdr(l), car(l))) < length(l) IMPLIES
(FORALL(x):O(q_s(g_l(cdr(l),car(l))))(x) = O(g_l(cdr(l),car(l)))(x),
length(leq_l(cdr(l), car(l))) < length(l) IMPLIES
(FORALL(x):O(q_s(leq_l(cdr(l),car(l))))(x) = O(leq_l(cdr(l),car(l)))(x)
|- null?(l),
O(app(q_s(leq_l(cdr(l), car(l))),
  cons(car(l), q_s(g_l(cdr(l), car(l))))))(x) = O(l)(x)
```

After splitting the antecedents of previous sequent, and instantiating twice, several lemmas are needed. Among them (see Fig. 2): occ_of_app, same_occ_leq and same_occ_g. Similarly, the proof command (lemma) is used to apply both lemmas leq_elements_size and g_elements_size in the branches rooted by ($\vdash_3$) and ($\vdash_4$). This is done to prove correct instantiation of the inductive hypotheses with lists that are in fact shorter than l. The lemma occ_of_app states that the occurrences of some element in an append is the sum of the occurrences in the appended lists and, after proper instantiations, adds the following antecedent to the previous sequent (node $\vdash_5$ in Fig. 2):

```
O(app(q_s(leq_l(cdr(l),car(l))),cons(car(l),q_s(g_l(cdr(l),car(l))))))(x)
=     O(q_s(leq_l(cdr(l), car(l))))(x) +
      O(cons(car(l), q_s(g_l(cdr(l), car(l)))))(x)
```

Then, the left-hand side of this equation is replaced by the right-hand side in the succedent and the second occurrence of O in the succedent is expanded. Since the expanded definition considers whether car(l) = x or not, the development of this proof brings the opportunity to relate (split) proof command and ($R_\wedge$) rule for the analysis of cases. These cases, car(l) = x and NOT car(l) = x (nodes $\vdash_6$ and and $\vdash_7$ in Fig. 2) are concluded by applying lemmas same_occ_leq and same_occ_g. The former lemma states that for an element k that is less than or equal to x, its occurrences in the lists l and leq_l(l,x) are equal, while it occurs zero times in the list g_l(l,x). The later one states that for an element k that is greater than x, its occurrences in the lists l and g_l(l,x) are equal, while it occurs zero times in the list leq_l(l,x).

5   Related Work and Conclusions

Teaching and motivating students to learn logic is, in general, a difficult task that has been gathering effort for a lot of years to ease the students learning and interest, giving rise to various tools aiming to achieve this goal. These tools range from those used to help students to work with evaluation of formulas in predicate logic over a concrete model, such as Tarski's World [4], to more sophisticated ones used to assist and verify the correct construction of proofs. Several tools can be used specifically to teach logic in various manners, according to Huertas' survey, used in this discussion, that classifies these tools into five categories: Provers, Checkers, Assistants, E-tutors and E-tutorials [11]. E-tutors and tutorials are directly related to a course and its contents, as for the first category, the user is more passive and the proofs can be done in an automatic way, reproduced by the student or assisted. Therefore, we believe that among these tools, checkers and assistants provide the greatest learning benefits for the students.

Assistants provide a high level of interaction and allow students to build their own deductions while offering help and guidance through messages, hints and dialog interfaces. This is the case of JAPE [5], a tool to reproduce paper and pencil proofs and display them using ND (in Fitch style), or SC deduction trees, allowing the user to apply deduction rules in a simple manner. JAPE usefulness to teach logic was reporter in [1]. It provides a full environment to the user, allowing, for instance, the cut rule for SC and proof by contradiction in ND, but neither compositions with partial proofs nor the use of equality rules. The use of partial proofs is supported by tools such as PANDA [10], a friendly graphic interface to deal with ND in classical logic. When the user clicks on a formula, the system provides a list of applicable rules, allowing backwards and forwards reasoning, displaying the proof tree as the user goes on with it in a very interactive way. Equality rules can also be found in other tools such as the *SC Trainer* in [9], a didactic tool that helps students to understand how to deal with SC. As another assistant PESCA can be also mentioned [18]. In PESCA the user can write down a logic formula in ASCII and then use the system to know which SC rule can be applied. The system does not support the use of cut rules or structural rules. The proofs are done step by step in a sequential manner, and the proof tree can then be constructed by user commands as a latex document. The corresponding ND proof can also be exported in a latex file.

Checkers can be used as a guide to students to check their deductions and verify if their exercises are correct. Here, ProofWeb [12] can be mentioned. Although it can be an assistant since it is a web interface where students can make use of proof assistants without installing them, this tool allows the introduction of deduction exercises that can be automatically verified when the students finish them. Other tools combine ProofWeb with other features in order to provide different environments to improve didactic, for instance, [21] presents an E-tutorial built to allow proving or refuting conjectures as a method to help teaching logic.

Even with so many tools available and the effort of several people in building more didactic and complete tools to ease the understanding of logical proofs, students can feel demotivated or unwilling to learn a subject if no real world contexts are presented that are related with the theory. In this sense, some logic classes focus on giving a general idea on the importance of this subject and its practical application, such as reported in [20], where examples of propositional and first-order logic sentences related to the day-to-day universe are provided and it is discussed how students can represent knowledge and reasoning about it, using for instance Prolog and other tools. This is valuable to relate the formal representation of real world sentences as logical propositions, being very helpful for teaching logic in philosophy courses, as reported in [19], where the ND Planner is used to teach ND to students with limited background in mathematics and scare familiarity with rigorous notations. However, when dealing with logic in CS courses, such tools still lack in providing an exciting practical use of logic in computational problems. Some tools provide this feature, such as VeriFun [23], a semi-automated tool allowing one to verify statements about programs written in a functional language, or even tools built over some proof assistant, such as the tool for imperative programs build over PVS discussed in [13].

Proof assistants have been shown valuable on logic classes that contain an extensive content, such as shown in [6,16,17], where Coq is used successfully to help motivating students to follow the necessary rigor when dealing with logic and programming languages by being a valuable guidance to the teacher, since the proof assistant is used as a "teaching assistant" in a large classroom. Another use of proof assistants to teach logic is done with a more friendly version of the ACL2, the ACL2 Sedan [8], that helps novices to use the tool in an "assisted" way by, for instance, preventing the user to make simple but common mistakes, such as inserting incomplete or not well-formed expressions. This version also aims to be self-teaching, by providing a graphical environment as an Eclipse plugin where the student can navigate and learn how to reasoning by programs with the assistant by using and exploring it.

Our proposal is to use proof assistants to provide a learning environment where the principles of logic can be related, in a general way, to reasoning about real computational problems. This is possible since proof assistants implement the required logic background and allow reasoning over CS and Math problems, by providing a specification language that allows one to deal with the semantics of programming and other computational features. Such tools are really powerful and thus applied to verify sophisticated properties, but also are adequate for our purposes since their basis is built just on formal deduction. Thus, the use of proof assistants can be done in a careful way, without demanding complex correctness proofs of the students, but just using the power provided by these tools to teach and motivate logic learning in a practical manner. The central premise of our proposal is to work first with very simple but real applications that motivate CS students to apply formal derivation as a non-fault approach in their programs. We use PVS to teach logic through stimulant and feasible applications in verification of simple functional programs. Although proof assistants are not designed as teaching tools, our goal is to show that a well-balanced use of them is useful for motivating students about the relevance and usefulness of formal deduction and induction in CS and Math.

This methodology has been applied to teach computational logic to undergraduate CS students with the background mentioned in the introduction and without previous experience with proof assistants. We believe our approach is pertinent and important to start the preparation of CS professionals, who will work with the construction and design of mathematically proved correct software. Since courses in proof theory, type theory and formal methods are optional, we advocate a long term, continuous and strong preparation as well as an early motivation of such professionals that should start during the first years of undergraduate CS courses.

In addition to formalization of algorithms, future work includes the use of proof assistants to teach complexity of algorithms, useful in courses on analysis of algorithms, as well as, to explore its natural application in courses on type theory as illustrated by Constable in [7].

References

1. Aczel, J., Fung, P., Bornat, R., Oliver, M., O'Shea, T., Sufrin, B.: Using computers to learn logic: undergraduates' experiences. In: Advanced Research in Computers and Communications in Education, pp. 875–882 (1999)
2. Almeida, A.A., Rocha-Oliveira, A.C., Ramos, T.M.F., de Moura, F.L.C., Ayala-Rincón, M.: sorting: a PVS Theory for sorting algorithms (2019). https://github.com/nasa/pvslib/tree/master/sorting - NASA Langley Research Center PVS libraries. Accessed Aug 2019
3. Ayala-Rincón, M., de Moura, F.L.C.: Applied Logic for Computer Scientists: Computational Deduction and Formal Proofs. UTCS. Springer, Cham (2017). https://doi.org/10.1007/978-3-319-51653-0
4. Barwise, J., Etchemendy, J.: The Language of First-Order Logic Including the Macintosh Version of Tarski's World 4.0. Monograph Collection (1993)
5. Bornat, R., Sufrin, B.: Jape: a calculator for animating proof-on-paper. In: McCune, W. (ed.) CADE 1997. LNCS, vol. 1249, pp. 412–415. Springer, Heidelberg (1997). https://doi.org/10.1007/3-540-63104-6_41
6. Chlipala, A.: Certified Programming with Dependent Types. MIT Press, Cambridge (2017)
7. Constable, R.L.: Formal systems, logics, and programs. In: Fitting, M., Rayman, B. (eds.) Raymond Smullyan on Self Reference. OCL, vol. 14, pp. 23–38. Springer, Cham (2017). https://doi.org/10.1007/978-3-319-68732-2_2
8. Dillinger, P.C., Manolios, P., Vroon, D., Moore, J.S.: ACL2s: The ACL2 Sedan. In 29th International Conference on Software Engineering ICSE, pp. 59–60. IEEE CS (2007)
9. Ehle, A., Hundeshagen, N., Lange, M.: The sequent calculus trainer - helping students to correctly construct proofs. In: Fourth Internation Conference on Tools for Teaching Logic TTL (2015). https://arxiv.org/abs/1507.03666
10. Gasquet, O., Schwarzentruber, F., Strecker, M.: PANDA: a proof assistant in natural deduction for all. A gentzen style proof assistant for undergraduate students. In: Blackburn, P., van Ditmarsch, H., Manzano, M., Soler-Toscano, F. (eds.) TICTTL 2011. LNCS (LNAI), vol. 6680, pp. 85–92. Springer, Heidelberg (2011). https://doi.org/10.1007/978-3-642-21350-2_11

11. Huertas, A.: Ten years of computer-based tutors for teaching logic 2000–2010: lessons learned. In: Blackburn, P., van Ditmarsch, H., Manzano, M., Soler-Toscano, F. (eds.) TICTTL 2011. LNCS (LNAI), vol. 6680, pp. 131–140. Springer, Heidelberg (2011). https://doi.org/10.1007/978-3-642-21350-2_16

12. Kaliszyk, C., Wiedijk, F., Hendriks, M., van Raamsdonk, F.: Teaching logic using a state-of-the-art proof assistant. In: Formal Methods in Computer Science Education FORMED, Satellite workshop of ETAPS 2008, pp. 111–120 (2008)

13. Lévy, M., Trilling, L.: A PVS-based approach for teaching constructing correct iterations. In: Wing, J.M., Woodcock, J., Davies, J. (eds.) FM 1999. LNCS, vol. 1709, pp. 1859–1860. Springer, Heidelberg (1999). https://doi.org/10.1007/3-540-48118-4_52

14. Owre, S., Rushby, J.M., Shankar, N.: PVS: a prototype verification system. In: Kapur, D. (ed.) CADE 1992. LNCS, vol. 607, pp. 748–752. Springer, Heidelberg (1992). https://doi.org/10.1007/3-540-55602-8_217

15. Owre, S., Shankar, N.: The formal semantics of PVS. Technical Report CR-1999-209321, NASA Scientific and Technical Information (STI) (1999)

16. Pierce, B.C.: Lambda, the Ultimate TA: Using a Proof Assistant to Teach Programming Language Foundations. In: SIGPLAN Not., pp. 121–122 (2009)

17. Pierce, B.C., et al.: Software Foundations. Electronic textbook (2014). https://softwarefoundations.cis.upenn.edu

18. Ranta, A.: PESCA - A Proof Editor for Sequent Calculus (2000). http://www.cse.chalmers.se/~aarne/old/pesca

19. Seligman, J., Thompson, D.: Teaching natural deduction in the right order with natural deduction planner. In: TTL (2015). https://arxiv.org/abs/1507.03681

20. Spichkova, M.: "Boring Formal Methods" or "Sherlock Holmes Deduction Methods"? In: Milazzo, P., Varró, D., Wimmer, M. (eds.) STAF 2016. LNCS, vol. 9946, pp. 242–252. Springer, Cham (2016). https://doi.org/10.1007/978-3-319-50230-4_18

21. Terrematte, P., Marcos, J.: TryLogic tutorial: an approach to Learning Logic by proving and refuting. TTL (2015). https://arxiv.org/abs/1507.03685

22. Troelstra, A.S., Schwichtenberg, H.: Basic Proof Theory. Cambridge UP, Cambridge (2000)

23. Walther, C., Schweitzer, S.: Verification in the classroom. J. Autom. Reasoning **32**(1), 35–73 (2004)

Teaching Formal Methods: From Software in the Small to Software in the Large

María-del-Mar Gallardo[(✉)] [iD] and Laura Panizo[iD]

Departamento de Lenguajes y Ciencias de la Computación, Universidad de Málaga,
Andalucía Tech, Campus de Teatinos s/n, 29071 Málaga, Spain
{gallardo,laurapanizo}@lcc.uma.es

Abstract. In this paper, we report the author's experience teaching formal methods to undergraduate students in the fourth year of the Software Engineering degree at the University of Málaga. The subject is divided into three blocks devoted to explaining the application of formal methods at different abstraction levels during the process of developing software. Although we teach the theoretical basis for students to understand the techniques, we mainly focus on the practical application of formal methods. Students are asked to realize in pairs three modelling and specifying projects of medium size (one for each block). The practical work corresponds to 60% of the student assessment, the remaining 40% is assessed with an exam on the theory of the subject matter. We have been teaching the subject during the last five years with very good results.

1 Introduction

The introduction of formal methods in the Computer Science curriculum is not an easy task. Many colleagues of other software engineering related areas tend to think that formal methods are only academic procedures without a real implementation in industry and that, consequently, they should be taught in master's degrees or courses for graduate students. However, the situation is evidently a vicious circle, if software engineers do not know formal methods, it is very unlike that they can be applied to industry. The change of Spanish university degree curricula in 2011 made it possible to update the content and structure of Computer Science studies. In the particular case of the University of Málaga, this change allowed the introduction of a subject devoted to Formal Methods in the fourth year of the Software Engineering degree with the consensus of a large part of the university community. This was possible probably because there are very established research groups with a strong background in Formal Methods in the Málaga University Computer Science Department.

This work has been supported by the Spanish Ministry of Science, Innovation and Universities project RTI2018-099777-B-I00 and the European Union's Horizon 2020 research and innovation programme under grant agreement No. 815178 (5GENESIS).

B. Dongol et al. (Eds.): FMTea 2019, LNCS 11758, pp. 97–110, 2019.
https://doi.org/10.1007/978-3-030-32441-4_7

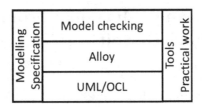

Fig. 1. Structure of the subject

In this paper, we describe our experience since 2014 teaching Formal Methods to groups of about 50 students for whom the subject is compulsory. Some colleagues think that formal methods are hard to teach and study due to their strong mathematical basis. However, our students' results show exactly the opposite. In general, students have enough skills in mathematics and logic to enable them to handle the subject very successfully.

Figure 1 shows the structure of the course which is taught during a semestre. Two transversal activities are present during the whole semestre. On the one hand, formal methods mainly consist of modelling and specification tasks. Thus, students must learn how to describe the software/hardware artifacts to be analyzed using different modelling languages. They must also know different logic-based specification languages to write the desirable properties. On the other, the use of (semi-)automatic tools that carry out the analysis work is the best way of convincing students of the usefulness of formal methods. Thus, during the course students become familiar with three different formal method tools.

As seen in Fig. 1, the subject is divided into three independent blocks. The idea is to show students that each formal technique is suitable for analyzing certain types of properties, from the fine and tiny errors produced by a bad coding of a *small* process or by the incorrect interaction of small processes to the incorrect construction of *large* software components whose inter-relations are not sufficiently well described. Thus, we start by introducing the foundations of explicit model checking. This method is very good at searching for errors produced by the incorrect interaction of concurrent processes. Model checking is a powerful technique capable of discovering subtle errors very difficult to find by non exhaustive methods. In consequence, we could say that model checking is a technique to analyze software in the small, that is, software written at a low-level where a simple instruction can cause system malfunction. Then, we continue with the ALLOY language and tool. This formal method contrasts with model checking since the input modelling language is declarative and the type of analysis carried out by Alloy is t a higher level, as it is mainly oriented to check structural properties. Finally, following this ascending path, the last formal method studied is UML/OCL. This is almost a must-choice for the software engineering students that use UML intensively during their degree. Thus, we show students that the use of a logic-based language like OCL is very useful to improve their UML descriptions which, in many cases, lack expressiveness.

W1	W2	W3	W4	W5	W6	W7	W8	W9	W10	W11	W12	W13	W14	W15
Model Checking						Alloy					OCL			

Fig. 2. Temporal distribution

The choice of the techniques taught in the course is guided by the research work carried out by several instructors in the Málaga University Computer Science department. Model checking is well known by the authors of this paper, and UML/OCL is the research core of other department members. Alloy was chosen as a link between model checking and UML/OCL. We were not specialist in the language. But after having worked with Alloy during the last few years, we think that it was a successful choice. Students like Alloy and the transition to OCL from Alloy is quite easy for them.

The paper is organized as follows. Section 2 summarizes the contents explained in each of the three blocks mentioned above. Section 3 explains the methodology used in the subject. Section 4 shows the assessment procedures and their results during the last five years. Finally, Sect. 5 gives the conclusions and discusses several aspects related to the results of the subject.

2   Content of the Subject

In this section, we give a brief review of the specific content of each subject block. The course begins with the usual introductory class that motivates the use of formal methods in software engineering. We show the serious consequences of software errors with examples such as the explosion of the Ariane 5 launcher[1] or the failure in the Patriot Missile[2]. They highlight the need of using formal methods not only in academia but also in industry to ensure that software behaves correctly, at least, with respect to its essential properties. In addition, in this introduction we explain the main activities of formal methods, i. e., modelling, specification, and analysis.

Then, we continue with the blocks devoted to the different formal methods. In each block, we describe the foundations of each formal technique and the corresponding tool. Thus, at the end of the course the students know how to model a system and specify their requirements with different languages, as commented in the Introduction. The rest of the section is devoted to presenting the main concepts and examples used in each of the following blocks:

1. Model Checking using Spin [7].
2. Satisfiability (SAT) with Alloy [8].
3. Testing of UML/OCL models with Use [5].

The temporal distribution of the blocks during the fifteen weeks of the course is shown in Fig. 2. As shown, the course devotes six weeks to Model Checking, five

---

[1] http://www-users.math.umn.edu/~arnold/disasters/ariane.html.
[2] http://www.ima.umn.edu/~arnold/disasters/patriot.html.

to ALLOY and four to OCL. The first block is longer because at the beginning of the course, students have to get used to the formal presentation inherent to formal methods and, also, because tool SPIN is a bit harder to handle. ALLOY is explained in approximately five weeks, and OCL only ocuppies four weeks.

2.1   Model Checking/SPIN

The first part of the subject focuses on model checking and the well-known tool SPIN [7]. We present the technique's main characteristics, and both its strengths (automatic technique, ability to produce counterexamples, etc.) and weaknesses (e.g. the state space explosion problem) are emphasized. The theoretical documentation given to the students (in the form of slides) is mainly extracted from [1]. The theoretical lectures and the presentation of the tool are interleaved in order to ease the practical work. The content of the block is organized to answer the following questions:

1. *How to model a system?* We initially introduce basic notions such as transition systems, reachability graphs, interleaving and synchronous/asynchronous composition of processes, and so on. All these concepts are illustrated with different practical examples described in natural language, graphical representation using state/transition diagrams, as well as mathematical notation. We use classical algorithms in this block, such as Peterson's mutual exclusion algorithm, the bit alternate protocol or the Sieve of Eratosthenes to show the communication along a pipeline of processes. After providing the theoretical background, we continue introducing the PROMELA modelling language to which the previously presented transition systems diagrams can be easily translated. In addition, we introduce tool SPIN showing its basic functionality available through the graphical user interface: syntax check, random and interactive simulation, and the verification of assertions and detection of deadlocks. Observe that, at this moment, students can only perform simulations and analysis of deadlock and assertion violation since temporal logic has not yet been explained.

2. *How to specify the system's requirements or properties?* In this chapter, we briefly introduce Kripke structures and temporal logic, after which we focus on the Linear Time Logic (LTL), which is one of the specification languages of properties accepted by SPIN. Understanding how to describe system requirements in LTL is one of the most challenging tasks of the course. We present the syntax and semantics of the boolean and temporal operators. Then, different kind of properties are introduced (e.g. invariants and liveness, safety and fairness properties) along with their LTL specification. At this point, we review the systems previously modelled in PROMELA to identify their main requirements and specify them using LTL. Students can now use SPIN to verify these requirements. They learn how to interpret the results of the analysis, and how to run and explore the counterexamples provided by the tool using the simulation features.

3. *What are the underlying algorithms to verify the system against the set of properties?* Finally, we present the notion of Büchi automata as the underlying representation of the system's behaviour and requirements in automata-based model checking algorithms. We also show the relation between Büchi automata and the transition system and LTL formulae. SPIN can accept requirements specified in LTL or directly described with a special PROMELA proctype called *never claim*. This proctype, is in fact, a Büchi automata whose execution is synchronously interleaved with the system's model. This part of the block is explained to show students the underlying algorithmic character of model checking, but we do not ask them to write Büchi automata.

As commented above, during the presentation of the theoretical aspects of model checking, students are asked to solve several practical exercises that go from the system modelling and the specification of requirements using LTL to its verification with SPIN, interpreting the tool output.

2.2   Alloy

The second block of the course is devoted to ALLOY [8], which is a tool and a language to model and specify systems based on sets and set relations. The tool, also called ALLOY analyzer, transforms the ALLOY model into an SAT problem that can be analyzed using different theorem provers, by default the tool uses the SAT4J prover. ALLOY limits the size of the system in order to achieve a compromise between automatic analysis and ease of use (normally not present in theorem provers). However, this is not a real limitation, since normally system errors can be discovered in small size systems.

The slides with the theoretical content about the language are based on the material of the ALLOY book [8]. This second block starts with a review of the theory of sets and relations. Although these concepts have been studied in previous courses, it is good to refresh them before introducing the ALLOY language. Although the syntax of ALLOY is not difficult, it is very extensive. This is why we follow an iterative approach to gradually introduce all the elements of the language, allowing students to do practical exercises almost from the beginning of the block. We carry out four iterations described in the following paragraphs.

1. In the first iteration, we introduce the main characteristics of the language and the tool as well as the basic constructors of the ALLOY language, that is, signatures and relations. In addition, we present multiplicities (for signatures and relations), set operators (e.g. union, intersection and difference), logic operators and relational operators. One of the most important and challenging operators is the composition operator *dot join* (.) that performs the composition of relations. This operator is a bit complicated to manage since relations can be composed from left to right (which is how the function composition usually works) but also from right to left. This is somehow counterintuitive but, on many occasions, it is very useful to simplify the specifications.

Finally, we introduce the concept of *facts* as the way to define system constraints. With all these elements, students are able to model several systems, such as an address book, and generate valid instances of the model using the ALLOY analyzer.

2. In the second iteration, we delve into the language details. For instance, we present abstract signatures and some new relational operators such as the transitive (^) and reflexive (*) closures, the transpose ($\sim$) operator, and the *let* and cardinality (#) operators. We also explain how to generate sets by comprehension, which is a powerful and compact way to define derived relations. All these new language elements are illustrated by means of several examples such as the classic family model which is built as a set of people with the father/child binary relations and their derived relations ancestor/descendant.

3. The third iteration introduces all the remaining language constructors needed to build the so-called static view of a system. Thus, students may already construct a complete system with multiple signatures, facts and predicates. We use the song "I'm my own grandpa", written by Dwight Latham and Moe Jaffe, given in [8], as an example of a system described in natural language, that can be translated step by step into ALLOY. In addition, in this iteration, we present the two types of analysis supported by ALLOY: running predicates to generate valid instances according to the model described, and checking assertion to search for counterexample instances.

4. The fourth iteration is devoted to describing how to construct dynamic ALLOY models by means of the inclusion of a new signature Time in the static models. Each instance of a static model can be seen as a snapshot of a system at a given time instant. To make models evolve over time, we use transitions that change the system state between to successive time instant. Transitions are implemented using ALLOY predicates. They must include three types of conditions: pre-conditions, that is, the state of the system before executing the transition, post-conditions, that is, the state of the system after executing the transition, and what remains unchanged, called frame conditions. In addition, we explain how to describe the initial system state and how to obtain an execution trace with ALLOY. To exemplify all these machinery, we extend the family model presented in the previous iterations with new predicates such as marriage or divorce, that can change the family relationships over time.

2.3   UML/OCL

The last block of the course is devoted to the refinement of UML descriptions using the OCL logic and the tool USE. Students know UML very well since they have already studied it in a previous course devoted to software modelling. Thus, students usually have a solid background in UML and we only have to refresh it with some examples. In fact, they also have some experience with OCL, although it is very limited. Thus, the block mainly focuses on presenting the subset of OCL supported by tool USE. OCL is an annotation-based language geared at the definition of constraints on UML class diagrams. Constraints are usually nested in the context of UML classes. For instance, OCL invariants impose constraints over the objects of a class, pre- and postconditions define the conditions that must be true before and after executing an instance method.

We start reviewing types and values, collections (set, bag and sequence) and the meta-types *OclAny* and *OclType* presented in OCL. Then, we continue presenting the navigation operator, which is the most characteristic of OCL to access remote objects. Navigation is especially tricky when it returns a collection of objects instead of a single one. To correctly navigate over collections, operations over collections have to be introduced. For instance *size()*, *isEmpty()*, or *notEmpty()* are related to the number of objects in a collection. Operations *select(B)* and *reject(B)* return the elements of a collection that satisfy or do not satisfy the condition *B*, respectively. Then, we also introduce OCL quantifiers over collections such as *forAll(B)*, the universal quantifier operator that returns *true* iff all elements in the collection satisfy boolean expression *B*, *exists(B)*, the existential quantifier that returns *true* iff there exists at least one element in the collection satisfying *B*, or *one(B)* that returns *true* iff there is exactly one element satisfying the condition *B*. We also use other operations such as *any(B)* that returns any element of the collection satisfying *B*. With all these operators, it is possible to define many invariants on classes.

This block includes many examples, and we usually present new ones every year. One of these examples, extracted from the OCL documentation, is a company's meeting system. Initially, the system comprises three classes `Meeting`, `TeamMember` and `Location`. Although the system's UML class diagram is simple, it is very useful to illustrate multiple invariants. For instance a meeting can impose a maximum number of attendees, or a meeting cannot take place in some locations. The complexity of the UML models is gradually increased, as well as the complexity of the corresponding OCL constraints.

We also introduce OCL pre- and post-conditions on methods. To define a postconditions it is useful to know the keywords *result*, which represents the result of a method, *@pre* to reference the value of an expression before executing a method, and operator send () that indicates that communication took place using the corresponding messages. To illustrate the definition of pre- and post-conditions, we extend the example that has already worked on. For instance, the meeting system example may be enriched with new methods to confirm a meeting, obtain the duration of a meeting, or change its date.

3   Methodology

In this section, we explain the methodology followed in the subject. The goal of the course is that students know the fundamental methodological basis of (some) formal methods and also learn that formal methods can help them during their carrier to develop complex and correct software. Thus, even though an approach to learning formal methods could entail emphasizing on their theoretical aspects, we discard this focus from the beginning. The course has about 50 students that are required to take it. Clearly, not all of them have good abilities to manage the complex formalizations involved in formal methods. It is worth mentioning, however, that every year we have found some students with extraordinary skills for formalization and abstraction, although that is not the normal case. In consequence, we decided to present formal methods with a strong practical bias. This means that each course block starts with some classes devoted to giving the basic theoretical notions as class lectures, following which we alternate theoretical lessons in the classroom and practice classes in the laboratory.

Each block has a list of practical exercises to help students become familiar with both the tool and the corresponding modelling and specification languages. Laboratory rooms have about 30 computers, so we need to use two of them (connected) for the practical sessions. Anyway, we encourage students to work in pairs, since it is easier for them to solve the problems collaboratively. Each block finishes with the so-called *assessable practical exercise*. This laboratory exercise involves modelling, specifying and analyzing a medium size problem. This practical work should be carried in pairs, and we use two or three laboratory sessions to help students get the job done. Anyway, we establish a deadline date to upload the code solving the problem, and a brief report explaining the solution, showing how the tool has successfully proved the correctness of the specified properties.

The key part of this methodology is to select, for each formal method, suitable examples that are neither too easy nor too complicated. Thus, each practical exercise should involve some effort to motivate students to learn and use the main characteristics of each formal method, but it should not be so difficult that many students cannot solve it. For this purpose, we structure practical exercises in several parts (from the easier questions to the much harder ones). This allows us to gradually distinguish the average students from the excellent ones, passing students that have worked sufficiently on it.

Since the methodology utilized in each block is the same, students have to carry out and submit three *assessable practical exercises*: one using model checking/SPIN, another one using ALLOY and the last one using OCLand the USE tool. Since the assessable practical exercises have significant weight in their final marks, students are very motivated to participate and do the job correctly. In practice, laboratory sessions are usually full of students working in pairs on the proposed exercises which from our point of view, constitutes the subject's main goals: (1) modelling, specifying and correctly using the associated tools and (2) learning to work in collaboration with other colleagues.

We now detail some of the theoretical and practical issues for each block.

3.1   Model Checking

With respect to the practical work, we start explaining the use of SPIN and PROMELA with simple exercises to try the tool and the main characteristics of languages like non-deterministic `do` and `if` sentences, boolean expressions as a mechanism to synchronize processes, the `assert` instruction to evaluate conditions at certain code points, and so on. In the first practical classes, students only use SPIN's simulation capacity and the verification of deadlocks (invalid end states). We use [7] to elaborate the list of practical exercises. Another good example that shows the effectiveness of SPIN/PROMELA to describe and analyze concurrent software is the version of the *session initiation protocol* (SIP) given by Prof. Pamela Zave in [10]. The example allows us to illustrate how it is possible to directly transform a transition system diagram into a PROMELA program. When the logic LTL is introduced in the theoretical lectures, we also use it to analyze safety and liveness properties on the PROMELA models using SPIN's verification capacities . Students also learn how to interpret the tool output (number of states analyzed, errors found, and so on), and to simulate the counterexamples provided by the tool to debug the code or correct the property specification.

Over the years, we have proposed different *assessable practical exercises*. As an example, the model checking *assessable practical exercise* in the 2016/17 course was based on the *Village Telephone System* described in [2]. The scenario is a small village where people use their phones to talk with any other neighbour (they do not care who the neighbour is). We elaborate a first PROMELA version of the system in which the village people communicate with each other through synchronous channels (which simplifies the implementation). In the second version of the exercise, the neighbours communicate through bounded asynchronous channels. This version is more complicated since neighbour processes have to solve the conflict that arises when two of them are simultaneously trying to talk to each other. Students had to prove the same safety/liveness temporal properties on both versions. For instance, some properties are "deadlock freedom", "no neighbour is ever talking with himself/herself" or "if all the people are talking, then some phone will eventually be hung up".

3.2   Alloy

The first practical exercises to be done in the laboratory are inspired by the examples in the ALLOY book [8]. In these examples, we try to make emphasis in the ability of the language to describe structural characteristics of software and to refine them using the relational first-order logic. Students take some time to assimilate some ALLOY operators such as the composition of relations or the transitive closure, but after solving some examples, they become familiar with them. With respect to the logic, they have also problems with some connectors and quantifiers (the implication is usually a bit confusing for them). However, with practice, they finally understand and use them correctly.

As in the case of model checking, students work on different assessable practical exercises each year. In this case, our goal is to use the language to describe systems structurally more complex than the ones analyzed by model checking, but in which, in contrast, the interaction between processes is not the main issue. As an example of the type of ALLOY assessable practical exercise, in the 2016/17 course, we proposed the modelling and specification of *Software-Defined Networks* (SDNs). An SDN is a new network paradigm in which the control and data planes are decoupled. Thus, switches are programmable by a controller node that can dynamically decide to change the way in which they should deal with a certain type of data packet. The example was inspired by the research work carried out by our group [4,9]. SDNs are a good example for showing the power of ALLOY since they have different actors which relate to each other in a non-trivial way. For instance, SDNs have different types of nodes (hosts, switches, controller) that are connected through ports via links. They also have control and data packets, tables in the switches with the rules to be applied to forward data packets, and so on. We structured this assessable practical exercise in four phases, from the simpler structural definition of SDNs to the complex behaviour where different predicates can be used to move the packets through the network.

3.3   OCL

When students reach the last block of the course devoted to OCL, they have enough experience in both the specification of properties using the first-order relational logic and in the construction of UML models. In consequence, in this block, we start with a brief reminder of the main OCL operators giving some non-trivial examples of how UML must be decorated with OCL expressions to improve the precision during the modelling phase. The slides with the theoretical documentation have been written using references [3,6]. In addition, we have used them to elaborate the list of exercises and we have added some exercises from the ALLOY list for students to compare the similarities and differences between the two languages. It is worth noting that this block is slightly shorter than the other two, since the student experience with formal methods at this point enables them to assimilate concepts quickly. As mentioned before, in this block we use the tool USE for the practical work in the laboratory. The tool has recently incorporated new functionalities to automatically generate object instances according to the UML/OCL specifications following the ALLOY philosophy. Students learn how to configure the tool to benefit from these new USE extensions.

As an example of assessable practical exercise for this block, in the 2015/16 course, we proposed the construction of a circular rail system composed of tracks, track sections, stations and trains. The example allows the iterative construction of the model in different phases as in the case of ALLOY. In the last phase, trains should move along the tracks preserving several invariants and pre/post conditions in the methods. We elaborated this practical exercise from scratch without using any previous reference material.

4   Assessment and Results

In this section, we explain how the subject is assessed and the results of these last few years.

As mentioned above, the assessable practical exercises constitute 60% of the final mark. We also give students a theory exam to distinguish between the work performed by each student in the practical exercises, and to detect plagiarism. The theoretical exam consists of solving three exercises, one for each block. The model checking exercise is to construct the reachability graph of some

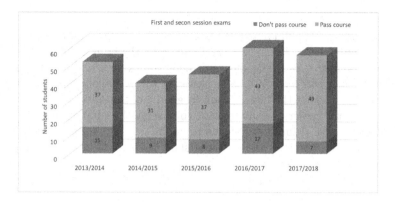

Fig. 3. Yearly results

small process models that interact via synchronous or asynchronous channels. We also ask students to specify some LTL properties on the model constructed. The ALLOY and OCL exercises are to implement small models with the aim of checking whether the score of the assessable practical exercise coincides with the student's abilities. In most cases, the results of the theoretical exam confirm the assessment of the practical exercises but, on some occasions there exists a significant discrepancy that shows that some of the students have not worked sufficiently on their practical exercises. Anyway, the following tables of results show that the methodology used in the subject makes it possible for students to learn formal methods and successfully pass the course.

Figure 3 shows the proportion between the students that pass/fail the subject after the first and second exam sessions. As shown, more than 70% of students pass each year.

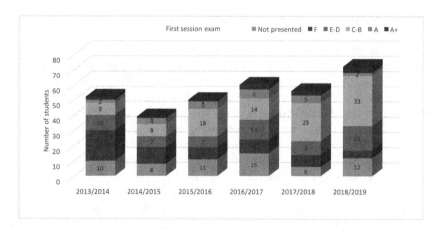

Fig. 4. Yearly detailed results

Figure 4 gives the results of the first exam session. For each year, it shows the number of students who did not take the final exam, who failed it and who passed it with different marks. The figure illustrates that the number of students with a very good performance in the subject is increasing year after year. In addition, we see that every year there are some students with excellent results.

Finally, Fig. 5 shows the distribution by gender of the subject. Although, it is not the goal of the paper, it is interesting to observe the small number of female students that course Computer Science in Spain.

From our point of view, these results show the success of the methodology followed in the subject.

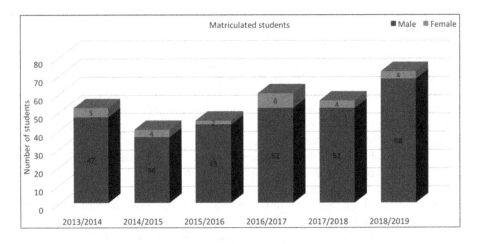

Fig. 5. Gender distribution

5   Conclusions

In this paper, we have described our experience teaching formal methods to undergraduate students at the University of Málaga. Since the subject is mandatory for all students in the fourth year of the Software Engineering degree, we have chosen to give it a very practical bias emphasizing the use of tools, although we also explain the theoretical basis using the usual terminology and formalism of Formal Methods. Thus, we have succeeded in having most students acquire skills in modelling, specifying and analyzing non-trivial software systems. We have chosen model checking with SPIN and OCL with the tool USE, since several research groups in our department are directly related to these two techniques. We think being very familiar with the methods we teach helps staying updated with the techniques and tools and transmitting enthusiasm to the students for the contents they are studying. ALLOY was chosen as an intermediate formal method between the exhaustive and automatic model checking technique and UML/OCL, whose analysis is currently less automated. Considering the results, we think it was a good choice, since ALLOY has proven to be an excellent technique that allows students to practice with a modelling language close to the object oriented languages they know very well and also with the first order logic to specify properties.

It is worth noting that specification using temporal logic and first-order logic is the hardest activity for the students. They are usually good at modelling, since they are very used to learning different programming languages of different paradigms (imperative, object-oriented and functional) and software tools are also very familiar to them. However, the logic specification of properties requires different nested logic operators (logic connectives, temporal operators and quantifiers) to be handled with ease. Some students find it difficult to use logic even though we show them many examples using well-defined formula patterns.

As a conclusion, we think that studying formal methods provides students with new and enriching tools to deal with software. On the one hand, formal methods allow them to look at software from a new perspective, improving their understanding and analysis abilities. On the other, students learn how formal method tools may help them improve the quality of the software developed, which can be of interest during their professional life.

From the teacher's point of view, the most complicated task is the elaboration of assessable practical exercises. We have to find an scenario that shows the power of formal methods to construct correct software (being attractive to students), and with a degree of complexity which allows us to fairly assess their competences in the subject. We think it would be a good idea to have a common space where the professors in charge of different formal method courses may share their practical projects.

To finalize, the course we have described here is geared at undergraduate students. We think that for master students, a formal method subject could be more theoretical although the practical perspective should always be present.

Acknowledgement. The authors would like to thank Dr. Francisco Durán and Dr. José María Álvarez for their help in designing the form and content of the course.

References

1. Baier, C., Katoen, J.P.: Principles of Model Checking (Representation and Mind Series). The MIT Press, Cambridge (2008)
2. Bhargavan, K., Gunter, C.A., Gunter, E.L., Jackson, M., Obradovic, D., Zave, P.: The village telephone system: a case study in formal software engineering. In: Grundy, J., Newey, M. (eds.) TPHOLs 1998. LNCS, vol. 1479, pp. 49–66. Springer, Heidelberg (1998). https://doi.org/10.1007/BFb0055129. http://dl.acm.org/citation.cfm?id=646525.694731
3. Cabot, J., Gogolla, M.: Object constraint language (OCL): a definitive guide. In: Bernardo, M., Cortellessa, V., Pierantonio, A. (eds.) SFM 2012. LNCS, vol. 7320, pp. 58–90. Springer, Heidelberg (2012). https://doi.org/10.1007/978-3-642-30982-3_3

4. Gallardo, M., Panizo, L.: Modelling and specifying software system with alloy (tutorial). In: Accepted to be presented in the Spanish Workshop on Programming Languages PROLE 2019 (Sistedes) (2019)
5. Gogolla, M., Büttner, F., Richters, M.: USE: a UML-based specification environment for validating UML and OCL. Sci. Comput. Program. **69**(1–3), 27–34 (2007)
6. Group, O.M.: OMG Object Constraint Language (2014). https://www.omg.org/spec/OCL/About-OCL/
7. Holzmann, G.: The SPIN Model Checker: Primer and Reference Manual. Addison-Wesley Professional, Reading (2003)
8. Jackson, D.: Software Abstractions: Logic, Language, and Analysis. The MIT Press, Cambridge (2006)
9. Lavado, L., Panizo, L., Gallardo, M., Merino, P.: A characterisation of verification tools for software defined networks. J. Reliable Intell. Environ. **3**(3), 189–207 (2017)
10. Zave, P.: Understanding SIP through model-checking. In: Schulzrinne, H., State, R., Niccolini, S. (eds.) IPTComm 2008. LNCS, vol. 5310, pp. 256–279. Springer, Heidelberg (2008). https://doi.org/10.1007/978-3-540-89054-6_13

On Teaching Applied Formal Methods in Aerospace Engineering

Kristin Yvonne Rozier(✉)

Iowa State University, Ames, IA, USA
kyrozier@iastate.edu

Abstract. As formal methods come into broad industrial use for verification of safety-critical hardware, software, and cyber-physical systems, there is an increasing need to teach practical skills in applying formal methods at both the undergraduate and graduate levels. In the aerospace industry, flight certification requirements like the FAA's DO-178B, DO-178C, DO-333, and DO-254, along with a series of high-profile accidents, have helped turn knowledge of formal methods into a desirable job skill for a wide range of engineering positions. We approach the question of verification from a safety-case perspective: the primary teaching goal is to impart students with the ability to look at a verification question and identify what formal methods are applicable, which tools are available, what the outputs from those tools will say about the system, and what they will not, e.g., what parts of the safety case need to be provided by other means. We overview the lectures, exercises, exams, and student projects in a mixed-level (undergraduate/graduate) Applied Formal Methods course (Additional materials are available on the course website: http://temporallogic.org/courses/AppliedFormalMethods/) taught in an Aerospace Engineering department. We highlight the approach, tools, and techniques aimed at imparting a good sense of both the state of the art and the state of the practice of formal methods in an effort to effectively prepare students headed for jobs in an increasingly formal world.

1 Introduction

Verification is a fundamental engineering skill; the current surge toward autonomy and increasingly intelligent operation of hardware, software, and cyber-physical systems has changed how we need to apply, and teach, verification at the university level. Industrial aerospace systems, including avionics, commercial aircraft, Unmanned Aerial Systems (UAS), satellites, and spacecraft, are being pushed toward design-for-verification, e.g., by Model-Driven Engineering [16,18,37–40], Fault Detection, Isolation and Recovery (FDIR) [5,14], and Runtime Verification [15,23,24,33]. "Nowadays, it is well-accepted that the development of critical [aerospace] systems involves the use of formal methods," [1].

Thanks to NSF CAREER Award CNS-1552934 for supporting this work.

B. Dongol et al. (Eds.): FMTea 2019, LNCS 11758, pp. 111–131, 2019.
https://doi.org/10.1007/978-3-030-32441-4_8

In addition to the obvious need to train verification engineers and researchers developing new formal methods, we are faced with the need to train a wide range of engineers with basic skills like understanding the outcome of a formal methods analysis.

Through a mixed-level (undergraduate/graduate) course, we introduce students to the fundamentals of formal methods, which we define as a set of mathematically rigorous techniques for the formal specification, design, validation, and verification of safety-critical systems, of which aircraft and spacecraft are the prime example. The course explores the tools, techniques, and applications of formal methods with an emphasis on real-world use-cases such as enabling autonomous operation. Students build experience in writing mathematically analyzable specifications from English operational concepts for real systems, such as aircraft and spacecraft. Together, the class examines the latest research to gain an understanding of the current state of the art, including the capabilities and limitations of formal methods in the design, verification, and system health management of today's complex systems. Students leave with a better understanding of real-world system specification, design, validation, and verification, including why the FAA specifically calls out formal methods in certification requirements such as DO-178B [21], DO-178C [20], DO-333 [19], and DO-254 [22].[1]

This course is intended to be a fun, interactive introduction to applying formal analysis in the context of real-world systems. We emphasize hands-on learning, through the use of software tools in homeworks and projects. Students learn the real tools used at NASA, Boeing, Collins Aerospace, Honeywell, Airbus, the Air Force, and others. Students from all areas of aerospace engineering, electrical and computer engineering, computer science, and other engineering disciplines, are encouraged to enroll. The course is cross-listed at the senior undergraduate/entry graduate levels and cross-listed in the Aerospace Engineering (AERE) and Computer Science (COMS) departments at Iowa State University and advertised in the Electrical and Computer Engineering and Mathematics departments; students from Industrial Engineering and Mechanical Engineering have also enrolled in this elective. Aiming for broad appeal, all concepts in the class are motivated chiefly through aerospace engineering applications; this shows direct applications to those students in the Aerospace Engineering Department and provides interesting use-cases for other majors. Example applications in homeworks include many different aspects of automated air traffic management and designing for autonomous operations of UAS.

Applied Formal Methods takes a *safety case* perspective [2, 7, 10, 16, 32]; in Aerospace Engineering, a safety case enables flight certification by providing an explicit statement of safety claims, a body of evidence concerning the system,

---

[1] Note that the railway industry has comparable standards CENELEC EN 50126 [8], EN 50128 [9], and EN 50129 [11]; these govern applications of formal methods in industrial rail systems, such as the success in verifying Paris' fully automatic, driverless Métro Line 14 (aka Météor-**Met**ro **e**st-**o**uest rapide) [3]. The course highlights railway, motor vehicle, medical, and other applications of industrial formal verification.

and an argument, based on the evidence, that the system satisfies its claims [31]. The major learning objective is for students to be able to read and understand, contribute to, and design an engineering system for being flight certified by a safety case, as this capability is now a general engineering skill. Students have the opportunity to construct a safety case as a half-semester final project for the course.

Learning Outcomes. Our central focus is to enable students to look at a problem, identify what we can verify, what information is needed to perform that analysis, how to validate the verification setup, and how to place the results in the field, e.g., by identifying what is now known, to what extent, and what is not known. Students learn to read research papers in formal methods, identify the current state of the practice, critically analyze current capabilities and limitations of the available tools and techniques, and effectively identify the inputs and outputs to verification, including what they really mean with respect to industrial safety standards. We specifically emphasize learning techniques for specification debugging and validation of mathematical models of systems. By the end of the course, students can identify what we can verify, and how; what can't we verify and why not; and what do we not have enough information to verify (and what additional information would we need). To construct an effective safety case, students must be able to recognize incomplete verification problems, identify ways to complete them, and identify assumptions and risks to validation.

Specific Learner Objectives. Through hands-on experience with formal methods tools and techniques, classroom discussions, homeworks, and projects students have the opportunity to learn to:

- Specify system requirements formally in Linear Temporal Logic (LTL) and Computational Tree Logic (CTL).
- Specify systems as formal models, i.e., models in a formal semantics.
- Apply model checking to system models and LTL specifications to determine if the models satisfy the specifications.
- Use tools popular in industrial verification labs, including explicit and symbolic model checkers, and theorem provers.
- Evaluate real-world systems to determine appropriate formal methods to use in their analysis.
- Evaluate system requirements, including determining if they are safety or liveness, and performing basic specification debugging.
- Analyze and draw conclusions about real-world systems regarding formal properties, understanding their significance and the inherent assumptions and limitations.
- Explain the principles underlying formal methods for different types of system analysis (e.g. design time versus runtime), the capabilities, and the limitations.
- Develop an understanding of the current state of the art and how to find formal methods tools for real-world use cases.

Prerequisites. The course requires the mathematical maturity and experience with proof structures covered in *Calculus II* (ISU MATH 166). Due to the cross-listing, the prerequisite is a disjunction of the Aerospace Engineering course *Computational Techniques for Aerospace Design* (ISU AERE 361) or the Computer Science *Algorithms* course (ISU COMS 311); both have MATH 166 as a prerequisite. Students should be familiar with first-order logic quantifiers and inductive proof techniques in order to understand Theorem Proving; professor permission enables registration for students who learned these skills in another 300-level course, e.g., from other engineering majors.

Organization. The remainder of this paper is organized as follows. Section 2 overviews the high-level approach to teaching Applied Formal Methods, including course assignments and examinations, highlights from the syllabus, and a general course schedule. We specifically pull out the tools and techniques covered in class in Sect. 3. Further details about the student research presentations and half-semester projects, including group projects, appear in Sects. 4 and 5 respectively. Section 6 concludes with an outlook toward continuous improvement of the course.

2   Approach

The first half (55–60%) of the course is a survey of the formal methods using modern tools, exemplified by case studies on industrial applications of formal verification. Class sessions are largely interactive and include discussions of the readings, guest speakers from industry, small group activities, and lecture. Students are encouraged to participate actively in class sessions. Lectures commence with "Formal Methods

Grade Component	Weight
Homeworks and Projects	30%
Midterm	25%
Research Paper Presentation	15%
Evaluation of Other Presentations	5%
Final Project	25%

Fig. 1. The weight assigned to each component: grades are assigned based on performance on homeworks, projects, presentations, a midterm exams, and a final project.

Explained: what are formal methods, why do we need formal methods, and why don't we formally verify everything?" The course proceeds to briefly review propositional logic and proofs. Class sessions cover in detail temporal logics, strategies for formal specification, specification debugging [27,28], system modeling, explicit model checking [29], theorem proving [6], and symbolic model checking based on [25]. These are the topics covered by the midterm examination, given during normal class hours, covering the material from readings and homeworks from the first half of the course. The second half of the course requires only two assignments: an in-class presentation of a research paper of the student's choosing, and a final project spanning the second half of the course, which serves in place of a final exam. All students are required to present the results of their final project mid-term and final results to the class and turn in a report including all artifacts required for reproducibility of their results at the end of the semester during the final exam period. Figure 1 summarizes the course assignments.

Other formal methods are discussed in class, included in in-class activities, demos or videos, and readings. Classes trace the relationships shown in Fig. 2.

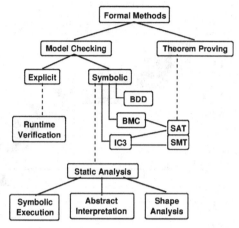

Homeworks and Projects. All homeworks are distributed and collected via github classroom. Homeworks are required to be typed and formatted in LATEX; some require submitting input files, e.g., for Spin, NUXMV, or PVS, solving the verification exercises. Students may talk about the problems with fellow students and the professor, but must submit individually-drafted write-ups. We occasionally discuss and work through parts of homework problems or variations thereof in class. When discussing

Fig. 2. The tree of Formal Methods, as presented in lectures; solid lines represent direct variations whereas dashed lines represent related derivations.

with fellow students they must strictly follow the "empty hands policy:" one cannot leave a discussion meeting with any record of the discussion (hard copy or electronic). All scratch paper must be torn and thrown away and all boards erased. Homeworks are encouraged to include BIBTEX references sections, including credit to collaborators and outside sources consulted. Students are encouraged to consult research papers, books, or other published materials in accordance with the University Honor Code (which prohibits searching for answers online, posting questions to internet forums, or discussing any assignments with others on the internet). All solutions should be written in each student's own words, even if the solutions exist in a publication referenced in the homework bibliography. While we adjust the course schedule every semester, depending on the students' backgrounds and the availability of guest speakers, a common schedule for the 16-week semester appears in Table 1.[2]

Reading Materials. Reading materials are included in the homeworks or otherwise distributed in class, e.g., research papers. There is no required textbook for this class. Two optional textbooks provide supplemental materials for students who desire additional reading, with the following caveats.

---

[2] In the U.S., there is usually a one-week break in the second half of the semester, after the mid-term project report presentations (Thanksgiving Break or Spring Break).

Table 1. A typical schedule for the homework assignments/small verification projects (top) comprising the survey of formal methods tools and techniques, along with the independent-research-based course assignments (bottom) across a 15-week semester with a following final exam period.

Week:	1	2	3	4	5	6	7	8	9	10	11	12	13	14	15	16	17
0: git	■																
1: PL	■																
2: TL		■															
3: Spec			■														
4: E-MC				■													
5: TP					■												
6: S-MC						■											
Midterm										■							
Pres									■	■	■	■	■	■	■		
P: Prop									■								
P: MP											■						
P: FP															■		
P: Fin																	■

HW 0	github classroom and LaTeXfundamentals; due 2nd class period
HW 1	Propositional Logic: review of logic and proof structure; ~1 week
HW 2	Temporal Logic: LTL and CTL; ~1.5 weeks
HW 3	Classifying Specifications & Explicit-State Modeling in Spin; ~2.5 weeks
HW 4	Explicit-State LTL Model Checking in Spin; ~1 week
HW 5	Theorem Proving: exercises in PVS (or Isabelle); ~1 week
HW 6	Symbolic Model Checking with NUXMV; ~1 week
Midterm	Comprehensive exam covering all homework topics in the 9th or 10th week
Presentation	Choice of paper due concurrently with HW 6; research paper presentation and peer evaluations during class periods after midterm
Project (P)	Initial project proposal due immediately following midterm; mid-way presentation (MP) in front of the class 2-3 weeks later; final presentation (FP) during the last week of classes; final paper/verification artifacts (Fin) due during final exam period

Optional Textbook: *An Introduction to Practical Formal Methods Using Temporal Logic* [13]. *Use this for:*

- good background on LTL: well-formed formulas, semantics, encoding English sentences, expressivity, normal forms, relationship to automata
- reactive system properties: safety, liveness, fairness
- specification and modeling of real systems
- deciding the truth of a temporal formula; related proof techniques including explicit model checking
- thorough chapter on Spin, including how to run it from the command line and a good Promela tutorial

- review of classical and propositional logic
- extensions including synthesizing software from specifications

Be cautious that:

- LTL is instead called PTL in this book; that is non-standard
- LTL2BA is not the best tool; SPOT is superior now: https://spot.lrde.epita. fr/
- URLs provided are outdated (no longer active or superseded by the state of the art)
- Spin chapter refers to outdated `xspin` (though only briefly)

Optional Textbook: *Systems and Software Verification: Model-Checking Techniques and Tools* [4]. *Use this for:*

- supplemental material on temporal logics (LTL, CTL, CTL*)
- background on automata as system models
- review of explicit and symbolic model checking
- reachability, safety, liveness, deadlock-freeness, fairness
- overview of modeling abstraction methods
- out-of-date chapters on SPIN and SMV still have useful reviews of basic tool usage
- ideas for related formal methods, including timed automata models, additional tools

Be cautious that:

- *This book is extremely out of date!*
- LTL is the proper name for Linear Temporal Logic (book calls it PLTL)
- comparisons of LTL vs CTL/CTL* have been changed/been disproved
- SMV version described is no longer available; current tool is NUXMV
- Spin version described has been updated (`xspin` vs `ispin`)

3 Tools and Techniques

While homeworks include hands-on projects in Spin, NUXMV, and PVS (or Isabelle), several other tools and techniques are covered in lectures, demos, or in-class activities. These tools, plus the most popular selections from student-devised projects, are collected in Table 2.

The first half of the semester (before the midterm) lectures are predominantly taught with a combination of slides and in-class exercises, frequently involving the class breaking into two or three groups, each with their own whiteboard, and solving problems in competition, usually in the form of a game. Groups must convince the rest of the class of the correctness of their answers to receive game points. The winning team is often awarded a prize like NASA stickers or similar swag from a guest speaker. Sometimes the same problem is posed to all groups, and sometimes each group is assigned a different strategy to employ then discuss with the class. For example, lessons on temporal logic encodings

Table 2. Tools featured in different areas of the Applied Formal Methods course.

	Spin Model Checker http://spinroot.com/ ✈ ✿
	SPOT Produces Our Traces https://spot.lrde.epita.fr/ ✿ ✿ (Optional for use in Spin-related homeworks)
	NuXmv Model Checker https://es-static.fbk.eu/tools/nuxmv/ ✈ ✿
	PVS Theorem Prover http://pvs.csl.sri.com/ ✈ (OR Isabelle) ✿
	Isabelle Theorem Prover https://isabelle.in.tum.de/ ✈ (OR PVS)
	PRISM Model Checker http://www.prismmodelchecker.org/ ✿ ✿
	Z3 SMT Solver https://github.com/Z3Prover/z3 ✿ ✿
	R2U2 Runtime Verifier http://temporallogic.org/research/R2U2/ ✿
	Dafny Language and Program Verifier http://rise4fun.com/dafny/ ✿

(*continued*)

Table 2. (*continued*)

	CBMC (Bounded Model Checker for C and C++ programs) http://www.cprover.org/cbmc/ ❂ ✿
	Coq Proof Assistant https://coq.inria.fr/ [Book: *Formal Reasoning About Programs* http://adam.chlipala.net/frap/] ❂ ✿

Legend:

✈	Required in homework assignments & covered thoroughly in class
❂	Featured in in-class instruction or presentation by guest lecturer(s)
✿	Utilized in student-selected final project(s)

involve dividing the class by their personal preferences into an LTL group, a CTL group, and an optional CTL* group (should anyone in the class feel most strongly about that logic). During this *Temporal Logic Showdown* (based in part on [35]), requirements in the form of English and/or figures (timelines, drawings, flowcharts, etc.) are posed to the class simultaneously. The first group to correctly encode the requirement in their logic and buzz in wins the points for the round. After that, encoding in the other logic (between LTL and CTL) earns half-points and the first team to buzz during that round in has the chance to steal those points by completing the correct encoding in the other team's logic and buzzing in before that team.

4   Research Paper Presentations

Each member of the class presents a research paper in applied formal methods to the class during the second half of the semester. A presentation consists of a slide presentation to the class covering the paper, and a discussion including the student's own analysis of its results. Students sign up for presentation times. The professor must approve all papers selected. Students can choose their papers from a provided list of papers or from a list of relevant publication venues. Alternatively, students may feel free to propose a paper on applying formal methods from any source for approval. Students evaluate the presentations of others for credit; anonymized summaries of the feedback of classmates are included in each student's evaluation. While the professor reads these evaluations, presentations are graded by the professor alone.

4.1 Professor's Presentation Evaluation Form

Students design their in-class research paper presentations according to the following evaluation criteria. Point values are listed in ⟦⟧s.

1. Did the presentation address the following aspects of the paper?
    (a) ⟦5⟧ What was the motivation given for the work? What problem was being solved or question was being answered?
    (b) ⟦5⟧ What was the product of the paper? How was the paper novel and what did it contribute to the field? What tools were used, problems were solved, and artifacts were created?
    (c) ⟦5⟧ Is the work in the paper *reproducible*,[3] i.e. are all of the necessary artifacts available to redo the study, including any models, specifications, theorems, code, data, benchmarks, or other instruments used to complete the study described in the paper.
    (d) ⟦5⟧ Is the work in the paper *correct*, i.e. did the authors specifically address how that they know their work is correct or provide any evidence of correctness such as a proof or a comparison to known results?
    (e) ⟦5⟧ Is the work in the paper *buildable*, i.e. is the foundation laid in such a way that others in the future would be able to build on it, extend it, and utilize the results in a meaningful way to accomplish a different project?
    (f) ⟦5⟧ Is there future work? This can include both future work listed in the paper and ideas the student has for extending the work.
2. ⟦10⟧ Did the presentation accurately overview the paper and the work presented therein, given the time limit? Did the student make an effort to fully understand the material and explain, if some piece is missing or not understandable, why that is the case?
3. ⟦20⟧ Was the presentation clear? Did the student make an effort to present the materials clearly and instructively, not necessarily in the order of the paper? Did the student draw on additional sources to fill out the information and background knowledge required to understand the paper? Did the student draw figures or create ways of presenting the material clearly and fully aside from simply pasting in artifacts from the paper?
4. ⟦15⟧ Did the student adequately cover background information and related work in an effort to enable him/herself as well as the class to understand the material being presented? Examples of doing this well might include the student reading and including material from some of the paper's citations or manuals for the tools used or otherwise including related background information to aid understanding of the material presented in the paper. These papers are short (usually about 15 pages) snapshots of single projects in formal methods and are meant to be read by practitioners familiar with the field and so usually do not include sufficient background information in the main text.

---

[3] For further reference on how exactly to define reproducibility, correctness, and buildability, please refer to: Rozier, Kristin Yvonne, and Rozier, Eric. "Reproducibility, Correctness, and Buildability: the Three Principles for Ethical Public Dissemination of Computer Science and Engineering Research," In IEEE International Symposium on Ethics in Engineering, Science, and Technology, Ethics'2014, May 23–24, 2014 [26].

4.2   Student's Presentation Evaluation Form

Peer evaluations earn students participation credit and provide good feedback that is summarized, anonymized, and returned to their peers. Point values are listed in [[]]s.

1. [[2]] What did you learn today? List at least three things you took away from today's class material.
2. [[1]] Was the presentation clear? What did you like about the way your classmate explained the materials to you? What constructive suggestions do you have to offer this classmate about how to present the material more clearly? (Your response will not be passed on to your classmate, however, an anonymized summary of all suggestions may be presented in class at the professor's discretion.)
3. [[1]] Was the content of the paper useful? Do you think the authors have contributed something that you or others might use or build upon in a future foray into formal verification? Why or why not?
4. [[1]] Is today's paper/formal method/topic something you think would be useful to examine in more depth in this class? Why or why not? Some paper topics may be covered in more depth following student presentations in the upcoming weeks; some may be earmarked for updating this class the next time it is taught.

5   Student Projects

In lieu of a final exam, students complete half-semester projects demonstrating their knowledge of applying formal methods. The high-level concept is simple: pick a system, pick a formal method, and successfully apply that method to that system. Students may work in groups of size one, two, or three. They are encouraged to discuss their proposal with the professor early and often; a formal project proposal is due mid-semester. Weekly progress reports, and a mid-term presentation to the class ensure steady progress while encouraging them to name their verification challenges and bring them up for discussion in class.

5.1   Initial Project Plan: Statement of Work

For the initial project plan, each person/group submits a statement of work that specifically addresses the following questions:

1. Define your group. Who are the members of your group? What is your group name?
2. Define the parameters of your project. What formal method are you using? What specifications will you verify? What system will you analyze?
3. What does a success look like for your project? For example, a successful model checking project will be able to demonstrate a system model, validation of that model, a set of temporal logic specifications, a set of model checking

runs checking the specifications against the model, and an analysis of the results. A successful theorem proving project will be able to demonstrate a set of (validated) theorems that automatically prove in an automated theorem prover and an analysis of the results of the proofs. A successful project in runtime monitoring will be able to demonstrate a set of specifications, a set of runtime monitors constructed from them, experimental results over many system runs demonstrating correct operation of the runtime monitors, and analysis of the results.

4. How will you demonstrate your analysis? In other words, answer all of the following questions that relate to your project:
   - What benchmarks will you use? Where will you get them from?
   - How will you demo your analysis (in the class?) (in your final report?)
   - How will you measure your results?

5. Remember to think about important logistics and organization elements. Each person/group will collaborate via a git repository shared with the professor. What will be the structure of your repo? How often should members check point models/specifications/documentation elements? If the project is a group project, how will the group coordinate? For a group, when will group meeting be? For a single-person project, what time have you scheduled each week to work on the project?

6. Provide a project timeline: for each week, list what you plan to accomplish that week. Be realistic and make backup plans! Your group will email the professor a (short) report at the end of each week with a project update according to your weekly plan. This email can be as simple as a statement that all tasks were accomplished that week, or as complicated as a detailed explanation why something did not work and how you have replanned to do an equivalent task. Weekly reports are due at 5pm on Fridays. This is your chance to get feedback on your progress and questions every week!

5.2 Progress Report and Preliminary Results

Provide a preliminary report from your group in the form of an in-class presentation of your results-so-far, making sure to **explicitly** answer the following questions:

- What parts of your project have you completed? Provide a bulleted list of work outputs to date.
- Provide an outline of your final report. What will the format be? What sections will you include? How do you plan to present any data and your analysis?
- What challenges have you encountered so far and how do you plan to overcome them? Provide a bulleted list of pairs {Challenge, Plan for action} to answer this section.
- Do you think you will need to change/modify/add to your project in any way? If so, make your case here. For example, if you have discovered that all of your specifications fail when analyzed against your system, what is your plan to modify the system and/or specifications?

5.3   Final Report and Presentation to the Class

Each person/group presents their project and results to the class during the last class periods, The time slots vary according to the size of the group. The final report from each person/group is due during the scheduled final exam period. The final report follows the outline and format described in the preliminary progress report. It includes the deliverables listed in the initial project plan/statement of work. Specifically, students should make sure to include the following:

- An abstract: succinctly summarize the final project setup and results.
- **All** models, specifications, code, or other artifacts needed to reproduce the work and re-run the verification tasks you completed for the project. **The professor must be able to re-run the verification procedure(s) followed.**
- Overview of the project including introduction, motivation, problem setup, and other information needed to understand the problem domain.
- Related work and background information, citing any resources used in the design and completion of this project.
- How was validation performed?
- What precisely was verified? What does it mean? How are the results significant?
- A complete verification analysis: results, performance of the tool(s) used, etc.
- A bibliography; Chicago Manual of Style (CMS) format is preferred.

The final report is cumulative; it needs to include *all* work done for the project in a complete report. Failure to include any of the required sections listed above results in losing points, even if the work was mentioned in class or in a presentation.

5.4   Example Student Projects

Students are encouraged to design final projects involving real-life systems of personal interest. Many students choose to form a project from the verification component of their graduate or undergraduate thesis research, or of a senior design or club project, such as creating a safety case for the launch of a student-designed CubeSat. Other popular categories of projects include designing tools to create instances of a game the student enjoys or to play such a game. Verification of autonomous driving or security scenarios from popular media, and "classic" projects (like verification of an elevator or traffic light protocol) have been proposed every semester. A competitive project category has emerged where two or three students all verify the same system from the same initial specification using a different favorite verification tool akin to an extended version of the VerifyThis[4] competition, with additional creative judging criteria.

---

[4] https://www.pm.inf.ethz.ch/research/verifythis.html.

Table 3 collects brief descriptions of student-designed final projects; in all cases, the size of the expected final deliverables scaled linearly with the number of students in the group and was adjusted for undergraduate vs graduate status. Several of the projects changed from the initial project proposal as the students ran into unexpected road blocks or discovered new tangents worth pursuing. Changes often stemmed from negative validation results, and ranged from minor adjustments in scope to major changes in the tools used (e.g., after being able to prove a construct could not be expressed in one tool), or problem objective. Accordingly, many of the final reports include thoroughly-explored negative results.

Table 3. A representative selection of student-devised final projects, 2015–2018.

Project description	#	U/G	Tool(s) used
Verify a lane-keeping module for autonomous cars. Starting with a road line detection algorithm, design a correct control algorithm, verify safety requirements using KeymaeraX and software implementation via CBMC, and validate including with real-world testing via augmenting the student's own car	1	U	KeymaeraX, CBMC
Utilize explicit model checking to generate 3×3 magic square puzzles with unique solutions, and to solve a given 3×3 puzzle	1	U	Spin
Analyze a real system (the CySat Make to Innovate (M:2:I) undergraduate research project) under active development spanning multiple abstraction layers on a demonstration mission toward surveying near-Earth objects under NASA's CubeSat Launch Initiative. Software and hardware verification that the ISU-designed flight computer meets mission reliability requirements	1	U	Spin, NuXmv
Verify the control of a tilt-wing medevac UAS designed by an ISU senior design team meets safety specifications	1	U	Spin
Generate attack graphs (structures representing all attack scenarios that an attacker can launch on a system) via a model-based approach with components/behaviors/defences/vulneratbilities and specification of security/resiliency properties. Iteratively model-check, disjuncting the previous counterexample to the current security property to generate acyclic attack graphs	1	G	AADL, Lustre, Jkind, AGREE
Model the ZigBee wireless protocol along with a collection of possible faults using OCRA for component based modeling, contract-based design and refinement, NuXmv for model checking of resulting transition systems, and xSAP for safety assessment and analysis	1	G	NuXmv, OCRA, xSAP
Use Spin to generate winning strategies for the Kartenspiele card game after a failed attempt with PVS	1	G	Spin, PVS

(*continued*)

Table 3. (*continued*)

Project description	#	U/G	Tool(s) used
Create a python library to parse mission-time linear temporal logic (MLTL), create an explicit state-space graph of a formula, display this with graphviz, and find a satisfying path through the graph, comparing two different search algorithms	1	G	N/A
Model a set of self-driving car intersection navigation scenarios and driving paths; use symbolic model checking to verify that the car always chooses a safe path. Generalize this to a maze solver, replicating previously-published experiments with TuLiP. Solve two small mazes using the GR(1)Py toolkit	1	G	NUXMV, TuLiP, GR(1)Py
Model, validate, and verify a set of traffic signaling algorithms using symbolic model checking. Scale the number of traffic lights to four per intersection and the number of successive intersections, varying properties like the timing of lights, max cumulative wait time, and max allowable queue length at a light. Compare performance for BDD, BMC, and IC3 back-ends	1	G	NUXMV
Define the formal operational semantics for a Simply Typed Message-passing Calculus (STMC) for software concurrency. Machine-checked proofs demonstrate the correctness of the message passing model including broadcasting, multicasting and guarded receive, and show the utility of the calculus by proving the properties guaranteed delivery of messages, the happens-before relation between the various actions, and the mover properties of the possible actions	1	G	Coq
Verify a vehicle-to-vehicle communication subsystem of an autonomous vehicle platooning system	1	G	Spin
Evaluate security of a Software Defined Network (SDN) model, including firewalls, a switch-level security feature to prevent malicious attacks, and a controller-level security feature to prevent DOS attacks by verifying invariants including reachability, isolation, loop freedom, no dead-ends	1	G	NUXMV
Formally analyze three security protocols Needham-Schroder Public Key Protocol, Otway- Rees Protocol and Kerberos Protocol. Analysis of a protocol is targeted towards detection of attacks in the protocol and suggestive modifications to the protocol that can eradicate the attack detected	1	G	NUXMV
Two students compete to verify the same Traffic Alert and Collision Avoidance System (TCAS) [34]: will explicit model checking or symbolic model checking be the better formal method for this task? One employs Holzmann's suggestions for optimizing the Spin model, the other takes advantage of NUXMV's newer back-end search algorithms. The competition includes performance, ease-of-use, modeling language expressibility, and usefulness of counterexamples	2	G	Spin, NUXMV

(*continued*)

Table 3. (*continued*)

Project description	#	U/G	Tool(s) used
Verify a python implementation of an A*-based pathfinding algorithm for a robot avoiding obstacles to traverse a maze via a shortest path using Linear Temporal Logic MissiOn Planning (LTLMoP). Validation included representing the same model in multiple tools and cross-validating model behaviors	2	U/G	(Py)NuSMV, PRISM, LTLMoP
Solve chess puzzles (puzzles over the pieces and rules of chess) via model checking focusing first on the mate-in-one-move problem	2	G	Spin
Verify a Mars rover mission sequence including coordination of a launch vehicle, ejection of the rover, executing a landing sequence, and commencing ground operations; confirm that mission goals are upheld including when faults occur and mitigation plans are executed	2	G	NUXMV
Explore the level of privacy maintained by users despite datamining, first through replication of a study on formal verification of privacy constraints on loan applications, then by devising a scalable model of e-voting machine data with user-specified privacy settings. An unsuccessful venture in Coq was followed by a successful re-imagining of the project using NUXMV	3	U^2/G	NUXMV, Coq
Compositionally verify an autonomous drone racing system with dissimilar components: localization (PVS), path planning (mCRL2). and the high-level architecture (Belief-Desire-Intent programming in AgentSpeak using Jason, Spin). Each student leads the verification of one subcomponent; ultimately the effort was unsuccessful due to integration challenges	3	U/G^2	PVS, mCRL2, Spin
Three students compete using three different tools to solve the same verification challenge (a Rubik's cube) and compare their results, performance, and which parts of the problem were easier/harder with each tool; creative methods of cross-validation took advantage of overlap between tools, e.g., NUXMV and MiniSat. Models started with $2 \times 2 \times 2$ cubes and scaled the difficulty and size of the cubes	3	G	Spin, NUXMV, CBMC, MiniSat and CaDiCaL SAT solvers
Model and verify a realistic subsystem of UTM (UAS Traffic Management) for near mid-air collision (NMAC) avoidance based on [12,17,30,36]. Use NUXMV to verify preflight, enroute, and emergency situations; further explore properties of enroute (like probability of a route change to avoid an NMAC) using PRISM	3	G	NUXMV, PRISM

Legend:

#	Number of students in the group
U	All students in group are undergraduate students
G	All students in group are graduate students

6   Conclusions and Outlook

In post-course surveys, students overwhelmingly identified details of tool use to be the aspect of the course they struggled with most; this includes the challenge of exposure to multiple new modeling/specification languages, details of tool installation/setup/debugging, and the gap between the level of detail required by formal methods tools versus their previous experiences, e.g., with pencil-and-paper proofs and informal (or no) system requirements. The majority of students identified the theorem proving tool (either PVS or Isabelle) as the most difficult to learn. When asked in hindsight (a year or more after course completion) what aspect(s) of the course turned out to be most useful, nearly every part of the course was listed by some student. The course project and the survey of formal methods were each identified by over half of the former students as most useful, citing in particular the perspective gained through experience. Other popular responses include the students' sound theoretical understanding of formal methods, the comparative discussions of specification languages, and in-class exercises (which some students felt so strongly about they questioned the ability to scale the course to include more students or online students). Several students particularly appreciated learning about the (ab)use of SAT solvers for a variety of applications including scheduling, specification debugging, and reduction of other problems to SAT. Nearly every student surveyed, both during the course and in hindsight, wrote an impassioned essay about the paper presentation section of the course, including the value of individualized feedback from the professor and other students, the opportunity to improve their analysis/presentation skills, exposure to the breadth of research frontiers and case studies in formal methods, the perspective they gained on verification in the wild, and the ability to steer the topics of the second half of the course to match the class' interests.

When asked how the course could be improved, students have overwhelmingly focused on small details of individual exercises; this feedback is continuously used to improve lectures, slides, and assignment descriptions. Examples include more in-class demonstrations of the quirkier aspects of tools, more details on industry standards requiring formal methods, and more information on community resources such as the active mailing lists for many tools, especially Isabelle and PVS. Students have requested add-on or follow-on courses such as a research paper reading group that offers an expanded version of the paper presentation portion of the class, and a large-scale application option where students work in groups to verify a real system over a whole semester simulating an industry setting. This is consistent with the most-requested course improvement: each semester students request more information on the end-to-end formal verification process, such as a universal flow-chart with all of the aspects of verification from initial conception to system maintenance laid out in fine detail.

Applied Formal Methods is currently taught as an elective; it counts toward one required technical elective for undergraduate and graduate students in Aerospace Engineering, Computer Science, and Computer Engineering, and has (so far) always been approved for replacing technical electives in other areas of engineering. Going forward, we look to integrate it as a required course in a track, e.g., in an avionics or intelligent systems concentration or minor within aerospace or in a cybersecurity or other interdisciplinary major. At its current size of 12–20 students per semester, the high level of participation and multiple presentations by each student in the course is both practical and advantageous: each student can participate actively in the course and receive personalized instruction in applying formal methods to a project tailored to her/his interests. Maintaining learning outcomes while potentially scaling the class to a larger size will be a formidable challenge. End-of-semester student ratings of the course have been consistently very high; if the course becomes required instead of purely elective, some adjustments may have to be made to accommodate a broader audience with more diverse interests.

One goal of publishing materials on the course is to receive feedback that can lead to continuous improvement; another is to open course materials for others to use and build upon. As formal methods teaching at the undergraduate and beginning graduate levels becomes more widespread there may be enough materials across the teaching community to support a tool-wise central repository of exercises, exam questions, and other teaching resources. We hope to contribute to such a repository, especially for tools like NUXMV and Spin, which remain popular for student use. Such materials could also be used to create industrial courses, such as the PVS Course at NASA Langley research center. We are continuously looking for industrial guest speakers to visit or give virtual lectures on their experiences applying formal methods in industrial practice. Traditionally, these lectures have received rave reviews and resulted in extra students showing up to class, in addition to those enrolled in the course. We hope to build up a club of regular industrial guest speakers as well as new lecturers to continue to inspire future students to apply formal methods in practice.

Acknowledgments. Information on our recent work can be found at: http://laboratory.temporallogic.org. Thanks to the Aerospace Engineering departments at Iowa State University and the University of Cincinnati for their forward thinking in recognizing the need to develop such a course. AERE/COMS 407/507 was developed over the Spring 2017, and Fall 2017 and 2018 semesters at ISU; parts of the class were first developed during the Spring 2015 and 2016 semesters at UC. Thanks to all of the students who actively participated in those courses, especially for coming up with such fantastic half-semester projects. Some course materials were inspired by or directly~derived from The TeachLogic Project (https://www.cs.rice.edu/~tlogic/); special thanks goes to Ian Barland, John Greiner, and Moshe Vardi for their brilliant teaching tools. Thanks to the NASA Langley Formal Methods Group for providing an excellent PVS course both in-person [6] and online with a rich collection of regularly-updated teaching materials. (https://shemesh.larc.nasa.gov/PVSClass2012/). Thanks to the many guest speakers including: Nikolaj Bjørner, Jonathan Hoffman, Yogananda Jeppu, César Muñoz, Lucas Wagner.

References

1. Ameur, Y.A., Boniol, F., Wiels, V.: Toward a wider use of formal methods for aerospace systems design and verification. Int. J. Softw. Tools Technol. Transf. **12**(1), 1–7 (2010)
2. Basir, N., Denney, E., Fischer, B.: Constructing a safety case for automatically generated code from formal program verification information. In: Harrison, M.D., Sujan, M.-A. (eds.) SAFECOMP 2008. LNCS, vol. 5219, pp. 249–262. Springer, Heidelberg (2008). https://doi.org/10.1007/978-3-540-87698-4_22
3. Behm, P., Benoit, P., Faivre, A., Meynadier, J.-M.: Météor: a successful application of B in a large project. In: Wing, J.M., Woodcock, J., Davies, J. (eds.) FM 1999. LNCS, vol. 1708, pp. 369–387. Springer, Heidelberg (1999). https://doi.org/10.1007/3-540-48119-2_22

4. Bérard, B., et al.: Systems and Software Verification: Model-checking Techniques and Tools. Springer, Heidelberg (2013). https://www.amazon.com/Systems-Software-Verification-Model-Checking-Techniques/dp/3642074782/ref=sr_1_1?ie=UTF8&qid=1483572091&sr=8-1&keywords=systems+and+software+verification

5. Bittner, B., et al.: An integrated process for FDIR design in aerospace. In: Ortmeier, F., Rauzy, A. (eds.) IMBSA 2014. LNCS, vol. 8822, pp. 82–95. Springer, Cham (2014). https://doi.org/10.1007/978-3-319-12214-4_7

6. Butler, R., et al.: NASA/NIA PVS Class 2012. NIA, Hampton, Virginia, USA, October 9–12 (2012). https://shemesh.larc.nasa.gov/PVSClass2012/online.html

7. Butler, R., Maddalon, J., Geser, A., Muñoz, C.: Simulation and verification I: formal analysis of air traffic management systems: the case of conflict resolution and recovery. In: Proceedings of the 35th Conference on Winter Simulation: Driving Innovation, pp. 906–914. Winter Simulation Conference (2003)

8. CENELEC, EN50126: Railway applications-the specification and demonstration of reliability. Availability, Maintainability and Safety (RAMS) (2001). https://www.cenelec.eu/standardsdevelopment/ourproducts/europeanstandards.html

9. CENELEC, EN50128: Railway applications-communication, signaling and processing systems-software for railway control and protection systems (2011). https://www.cenelec.eu/standardsdevelopment/ourproducts/europeanstandards.html

10. Denney, E., Pai, G., Pohl, J.: Heterogeneous aviation safety cases: integrating the formal and the non-formal. In: 2012 IEEE 17th International Conference on Engineering of Complex Computer Systems, pp. 199–208. IEEE (2012)

11. EN50129, CENELEC: Railway applications-communication, signalling and processing systems-safety related electronic systems for signalling. British Standards Institution, United Kingdom. ISBN, pp. 0580–4181 (2003)

12. von Essen, C., Giannakopoulou, D.: Analyzing the next generation airborne collision avoidance system. In: Ábrahám, E., Havelund, K. (eds.) TACAS 2014. LNCS, vol. 8413, pp. 620–635. Springer, Heidelberg (2014). https://doi.org/10.1007/978-3-642-54862-8_54

13. Fisher, M.: An introduction to practical formal methods using temporal logic, vol. 82. Wiley Online Library (2011). https://www.amazon.com/Introduction-Practical-Formal-Methods-Temporal-ebook/dp/B005E8AID2/ref=sr_1_1?ie=UTF8&qid=1483648485&sr=8-1&keywords=practical+formal+methods+using+temporal+logic

14. Gario, M., Cimatti, A., Mattarei, C., Tonetta, S., Rozier, K.Y.: Model checking at scale: automated air traffic control design space exploration. In: Chaudhuri, S., Farzan, A. (eds.) CAV 2016. LNCS, vol. 9780, pp. 3–22. Springer, Cham (2016). https://doi.org/10.1007/978-3-319-41540-6_1

15. Geist, J., Rozier, K.Y., Schumann, J.: Runtime observer pairs and bayesian network reasoners on-board FPGAs: flight-certifiable system health management for embedded systems. In: Bonakdarpour, B., Smolka, S.A. (eds.) RV 2014. LNCS, vol. 8734, pp. 215–230. Springer, Cham (2014). https://doi.org/10.1007/978-3-319-11164-3_18

16. Guarro, S., et al.: Formal framework and models for validation and verification of software-intensive aerospace systems. In: AIAA Information Systems-AIAA Infotech@ Aerospace, p. 0418 (2017)

17. Kochenderfer, M.J., Chryssanthacopoulos, J.: Robust airborne collision avoidance through dynamic programming. Massachusetts Institute of Technology, Lincoln Laboratory, Project Report ATC-371 (2011)

18. Mattarei, C., Cimatti, A., Gario, M., Tonetta, S., Rozier, K.Y.: Comparing different functional allocations in automated air traffic control design. In: Proceedings of Formal Methods in Computer-Aided Design (FMCAD 2015), Austin, Texas, USA. IEEE/ACM, September 2015

19. Radio Technical Commission for Aeronautics: DO-333 – formal methods supplement to DO-178C and DO-278A (2011). https://www.rtca.org/content/standards-guidance-materials

20. Radio Technical Commission for Aeronautics: DO-178C/ED-12C – software considerations in airborne systems and equipment certification (2012). https://www.rtca.org/content/standards-guidance-materials

21. Radio Technical Commission for Aeronautics (RTCA): DO-178B: Software considerations in airborne systems and equipment certification, December 1992

22. Radio Technical Commission for Aeronautics (RTCA): DO-254: Design assurance guidance for airborne electronic hardware, April 2000

23. Reinbacher, T., Rozier, K.Y., Schumann, J.: Temporal-logic based runtime observer pairs for system health management of real-time systems. In: Ábrahám, E., Havelund, K. (eds.) TACAS 2014. LNCS, vol. 8413, pp. 357–372. Springer, Heidelberg (2014). https://doi.org/10.1007/978-3-642-54862-8_24

24. Rozier, K.Y., Schumann, J., Ippolito, C.: Intelligent hardware-enabled sensor and software safety and health management for autonomous UAS. Technical Memorandum NASA/TM-2015-218817, NASA, NASA Ames Research Center, Moffett Field, CA 94035, USA, May 2015

25. Rozier, K.: Linear temporal logic symbolic model checking. Comput. Sci. Rev. J. 5(2), 163–203 (2011). https://doi.org/10.1016/j.cosrev.2010.06.002

26. Rozier, K., Rozier, E.: Reproducibility, correctness, and buildability: the three principles for ethical public dissemination of computer science and engineering research. In: IEEE International Symposium on Ethics in Engineering, Science, and Technology, Ethics 2014, pp. 1–13. IEEE, May 2014

27. Rozier, K.Y., Vardi, M.Y.: LTL satisfiability checking. In: Bošnački, D., Edelkamp, S. (eds.) SPIN 2007. LNCS, vol. 4595, pp. 149–167. Springer, Heidelberg (2007). https://doi.org/10.1007/978-3-540-73370-6_11

28. Rozier, K.Y., Vardi, M.Y.: A multi-encoding approach for LTL symbolic satisfiability checking. In: Butler, M., Schulte, W. (eds.) FM 2011. LNCS, vol. 6664, pp. 417–431. Springer, Heidelberg (2011). https://doi.org/10.1007/978-3-642-21437-0_31

29. Rozier, K.Y., Vardi, M.Y.: Deterministic compilation of temporal safety properties in explicit state model checking. In: Biere, A., Nahir, A., Vos, T. (eds.) HVC 2012. LNCS, vol. 7857, pp. 243–259. Springer, Heidelberg (2013). https://doi.org/10.1007/978-3-642-39611-3_23

30. NASA UTM Research Transition Team (RTT): NASA UTM NextGen concept of operations v1.0, May 2018. https://utm.arc.nasa.gov/docs/2018-UTM-ConOps-v1.0.pdf

31. Rushby, J.: A safety-case approach for certifying adaptive systems. In: AIAA Infotech@ Aerospace Conference and AIAA Unmanned... Unlimited Conference, pp. 1–16 (2009)

32. Rushby, J.: Logic and epistemology in safety cases. In: Bitsch, F., Guiochet, J., Kaâniche, M. (eds.) SAFECOMP 2013. LNCS, vol. 8153, pp. 1–7. Springer, Heidelberg (2013). https://doi.org/10.1007/978-3-642-40793-2_1

33. Schumann, J., Moosbrugger, P., Rozier, K.Y.: R2U2: monitoring and diagnosis of security threats for unmanned aerial systems. In: Bartocci, E., Majumdar, R. (eds.) RV 2015. LNCS, vol. 9333, pp. 233–249. Springer, Cham (2015). https://doi.org/10.1007/978-3-319-23820-3_15

34. U.S. Department of Transportation Federal Aviation Administration: Introduction to TCAS II version 7.1, February 2011. hQ-111358. https://www.faa.gov/documentlibrary/media/advisory_circular/tcas%20ii%20v7.1%20intro%20booklet.pdf

35. Vardi, M.Y.: Branching vs. linear time: final showdown. In: Margaria, T., Yi, W. (eds.) TACAS 2001. LNCS, vol. 2031, pp. 1–22. Springer, Heidelberg (2001). https://doi.org/10.1007/3-540-45319-9_1

36. Wei, P., Atkins, E., Schnell, T., Rozier, K.Y., Hunter, G.: NSF PFI:BIC: pre-departure dynamic geofencing, en-route traffic alerting, emergency landing and contingency management for intelligent low-altitude airspace UAS traffic management, July 2017. https://www.nsf.gov/awardsearch/showAward?AWD_ID=1718420

37. Wiels, V., Delmas, R., Doose, D., Garoche, P.L., Cazin, J., Durrieu, G.: Formal verification of critical aerospace software. AerospaceLab (4), 1–8 (2012). https://hal.archives-ouvertes.fr/hal-01184099

38. Zhao, Y., Rozier, K.Y.: Formal specification and verification of a coordination protocol for an automated air traffic control system. In: Proceedings of the 12th International Workshop on Automated Verification of Critical Systems (AVoCS 2012). Electronic Communications of the EASST, vol. 53. European Association of Software Science and Technology (2012)

39. Zhao, Y., Rozier, K.Y.: Formal specification and verification of a coordination protocol for an automated air traffic control system. Sci. Comput. Program. J. 96(3), 337–353 (2014)

40. Zhao, Y., Rozier, K.Y.: Probabilistic model checking for comparative analysis of automated air traffic control systems. In: Proceedings of the 33rd IEEE/ACM International Conference On Computer-Aided Design (ICCAD 2014), San Jose, California, USA, pp. 690–695. IEEE/ACM, November 2014

Effective Teaching Techniques

Teaching Concurrency
with the Disappearing Formal Method

Emil Sekerinski$^{(\boxtimes)}$ ⓘ

McMaster University, Hamilton, ON, Canada
emil@mcmaster.ca
http://www.cas.mcmaster.ca/~emil

Abstract. The Gries-Owicki non-interference condition is fundamental to concurrent programming, but difficult to explain as it relies on proof outlines rather than only pre- and postconditions. This paper reports on teaching a practical course on concurrent programming using hierarchical state diagrams to visualize concurrent programs and argue for their correctness, including non-interference.

Keywords: Non-interference · Guarded commands · State diagrams

1 Introduction

Given the ubiquity of distribution, interactive computing, and multi-core processors, concurrency is pervasive. Concurrent programming is also error-prone. The intrinsic difficulty is succinctly captured by the Gries-Owicki rule for the non-interference of concurrent processes [17]. Let the statements of an abstract programming language be made up of assignments ("$:=$") that are composed sequentially ("$;$"), conditionally ("**if**"), and repetitively ("**while**"). Statements can also be composed in parallel ("$\|$"), be guarded ("**await**"), and put into atomicity brackets ("$\langle \ldots \rangle$"). Let S_i be statements such that, considered in isolation, each S_i under precondition P_i establishes postcondition Q_i, for $i \in 1..n$:

$$\{P_1\} \, S_1 \, \{Q_1\}, \; \ldots \, , \{P_n\} \, S_n \, \{Q_n\}$$

The Gries-Owicki rule states that under the conjunction of the preconditions P_i, the parallel composition of S_i establishes the conjunction of the postconditions Q_i,

$$\{P_1 \wedge \ldots \wedge P_n\} \, S_1 \, \| \, \ldots \, \| \, S_n \, \{Q_1 \wedge \ldots \wedge Q_n\}$$

provided that all atomic statements R in each S_i do not interfere with any annotation of S_j for all $i, j \in 1..n$ with $i \neq j$. That is, the processes S_i are not considered to be "black boxes" leading from a precondition to a postcondition,

Supported by NSERC Grant RGPIN-2017-06692.

B. Dongol et al. (Eds.): FMTea 2019, LNCS 11758, pp. 135–149, 2019.
https://doi.org/10.1007/978-3-030-32441-4_9

but composed of atomic statements with intermediate annotations. In general, an atomic statement R with precondition A and postcondition B,

$$\{A\}\ R\ \{B\}$$

does not interfere with condition C if R preserves C:

$$\{A \wedge C\}\ R\ \{C\}$$

Each process S_i has to "know" about the conditions in visible intermediate states of all other processes (*proof outline*) and has to ensure that its own actions (atomic statements) do not invalidate those conditions; these conditions can be over the global variables as well as the private ones of any involved processes. The necessity of knowing the private variables of other processes and the exploding number of non-interference checks arising from the combinations of intermediate states and atomic statements in all process explains the intrinsic difficulty in "getting a concurrent program right".

The above programming language does not correspond to commonly used programming languages, but allows to define constructs of those: compare-and-swap instructions of processors, semaphores, monitors, remote procedure calls, synchronous and asynchronous channels, and any other concurrency constructs that can be understood in terms of interleaving. This model of concurrency allows to motivate, compare, and contrast concurrency constructs in current languages. The non-interference condition allows to analyze restrictions on the structure of process and the shape of annotations to minimize interference [1,2,9]. It is therefore appealing for courses that involve practical programming. However, explaining non-interference with proof outlines requires a long introduction.

The author was teaching a sequence of two software design courses that include (sequential) programming with pre- and postconditions as a unifying foundation for requirements analysis, stepwise development, modularization, object-oriented programming, and testing [20]. The experience has been that the notions of pre- and postconditions, loop and class invariants are well received, but a rigorous application of the correctness rules not. The source of the difficulty are the programming languages to which students have been previously exposed: the confusion of assignment and equality due to the use of = for assignment (and still pronouncing it as "equal"), the confusion of statements and expressions, and to some degree, the use of {...} in programming languages for bracketing statements. Together with the distinction of **assert** statements and correctness assertions, this results in a thorough confusion of statements, properties of those, and the notations for each, as visible in typical nonsensical expressions of the kind:

$$\{true\}\ x := x + 1\ \{x = x + 1\} \tag{1}$$
$$\{x = 0\} \Rightarrow \{x \geq 0\} \tag{2}$$

(One can attribute the recent interest in functional programming to the ugliness of statements like $x = x + 1$. According to Kernighan and Ritchie [13, p. 17], the justification for using = for assignment is: "Since assignment is about twice as frequent as equality testing in typical C programs, it's appropriate that the operator be half as long". Later on, Ritchie comments on the development of C by "Other fiddles ... remain controversial, for example the decision to use the single character = for assignment instead of :=." [18]).

Defining formally the grammar of statements and expressions first, as Dijkstra [7], to avoid above nonsense had a limited effect: it distracts, it is perceived as an artificial restriction, and it does not help in understanding the difference between program statements, program expressions, and correctness assertions (and (1) is syntactically correct). Students were perfectly able to explain all these concepts correctly, but would still write nonsensical expressions. They are not to blame, having been exposed to confusing notations.

This paper reports on teaching formally concurrent programming with the non-interference condition to third-year students who had a prior course in discrete math and logic, following Gries and Schneider [10] and using Calc-Check [12], and limited exposure to pre- and postconditions. Assignments involve programming in Python (with semaphores), Java (with monitors), and Go (with message passing). As the novelty, students are introduced to nested, concurrent state diagrams: the graphical layout makes the distinction between statements and annotations obvious; it gives a simple visualization of the non-interference condition; it avoids curly braces, so does not compete with programming languages; it explains naturally nondeterminism and blocking; it is not perceived as a "formal method" with its own language but as a visual model of programs.

State diagrams with nesting and concurrency originate in Harel's statecharts for embedded systems [11]. Manna and Pnueli use nested states with invariants for verification [16]. Back proposes to start the development with nested invariants diagrams and then to add transitions [5]. By contrast, here we present invariants and transitions hand-in-hand, with the structure of transitions emerging from structured (single-entry, single-exit) programs. This can lead to different nested state structure, e.g. the final state of a simple loop is nested inside the invariant state in Back but a separate state here. State invariants here are similar to invariantcharts in our earlier work on embedded systems [19]. Invariantcharts have nesting, concurrency, events, and broadcasting like statecharts, as well as invariants, but we do not consider events and broadcasting here. Our contribution is a visual interpretation of the non-interference condition.

The next section outlines the course material on state diagrams for sequential programming, including correctness, hierarchical diagrams, nondeterminism, and blocking. Section 3 outlines the presentation of state diagrams with concurrency, atomicity, and non-interference and gives an example of using semaphores. Section 4 provides some specifics of the course delivery and Sect. 5 discusses the approach.

2   State Diagrams for Sequential Programming

The presentation of state diagrams below follows the exposition in class.

Basic Statements. Programs in our *algorithmic notation* consist of *variables* that hold values, *expressions* that, when evaluated, have a result, and *statements* that, when executed, have an effect on variables. The *assignment statement* x := E evaluates expression E and assigns the result to variable x. This can be generalized to *multiple assignments* in which two or more variables are modified, as in x, y := E, F. The sequential composition S; T first executes statement S and then statement T:

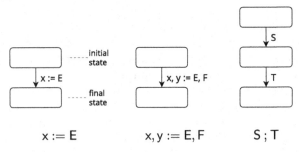

$$x := E \qquad x, y := E, F \qquad S; T$$

The *conditional statement* evaluates a Boolean expression and executes a statement or does nothing, depending on the evaluation of the expression. The repetitive statement executes the *body*, a statement, as long as the condition, a Boolean expression, is true. Suppose S, T are statements and B is a Boolean expression:

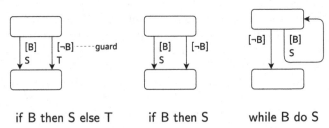

$$\textsf{if B then S else T} \qquad \textsf{if B then S} \qquad \textsf{while B do S}$$

Correctness. Statements have a number of properties; the property we consider here is which final state is produced for which initial state. More generally, we consider the final states for a set of possible initial states. The initial and final states are characterized by predicates. In general, for predicates (Boolean expression) P, Q, we express that under precondition P statement S establishes postcondition Q by a *correctness assertion* (colour distinguishes programs from properties):

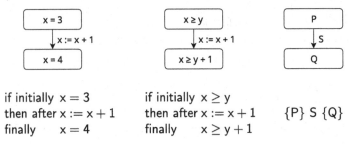

if initially $x = 3$ if initially $x \geq y$
then after $x := x + 1$ then after $x := x + 1$ $\{P\}\ S\ \{Q\}$
finally $x = 4$ finally $x \geq y + 1$

The correctness of a possibly composed statement with respect to its precondition, postcondition, and *annotations* in intermediate states is checked in state diagrams by following rules for transitions between states:

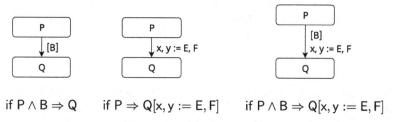

$$\text{if } P \wedge B \Rightarrow Q \qquad \text{if } P \Rightarrow Q[x, y := E, F] \qquad \text{if } P \wedge B \Rightarrow Q[x, y := E, F]$$

Here, $P[x, y := E, F]$ stands for the simultaneous substitution of x with E and y with F. A statement is correct if all its transitions are correct. For example, the algorithm for multiplication by successive additions is:

$$\{x \geq 0\}$$
$$z, u := 0, x$$
$$\{z + u \times y = x \times y \wedge u \geq 0\}$$
$$\text{while } u > 0 \text{ do}$$
$$\quad z, u := z + y, u - 1$$
$$\{z = x \times y\}$$

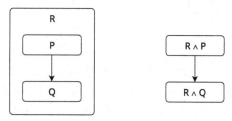

According to the correctness rules of transitions, the algorithm is correct provided:

1. $x \geq 0 \Rightarrow (z + u \times y = x \times y \wedge u \geq 0)[z, u := 0, x]$
2. $z + u \times y = x \times y \wedge u \geq 0 \wedge u > 0 \Rightarrow$
   $\qquad (z + u \times y = x \times y \wedge u \geq 0)[z, u := z + y, u - 1]$
3. $z + u \times y = x \times y \wedge u \geq 0 \wedge u \leq 0 \Rightarrow z = x \times y$

These follow from the rules of [10]. The precondition of the repetition has the role of an *invariant*.

Hierarchical Diagrams. Nesting allows an annotation that is repeated in several states to be factored out into a *superstate*. Following are equivalent diagrams:

As long as the computation resides within the superstate, R is an invariant. For example, the algorithm for multiplication by shifting is:

$$\{x \geq 0\}$$
$$z, u, v := 0, x, y$$
$$\{z + u \times v = x \times y \wedge u \geq 0\}$$
while $u \neq 0$ do
   if odd(u) then
      $z := z + v$
   $u, v := u$ div $2, 2 \times v$
$$\{z = x \times y\}$$

Following are two equivalent diagrams; the one to the right factors out $u > 0$ to a superstate:

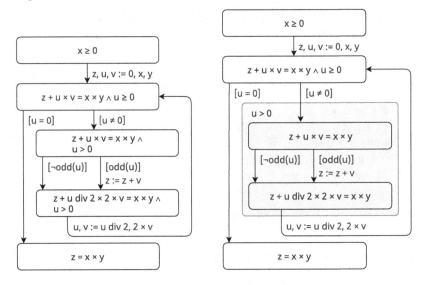

Nondeterminism and Blocking. Dijkstra's if and do *guarded commands* generalize the conditional and repetitive statements to multiple alternatives, each with its own guard. If several guards hold, one is selected nondeterministically. If none holds, the alternative stops and the repetition terminates:

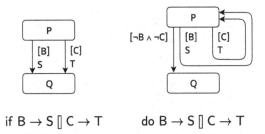

if $B \to S [] C \to T$ do $B \to S [] C \to T$

As special cases, $S [] T$ selects nondeterministically between S and T, as in an alternative with true guards, skip does nothing, and stop does not lead to any further state:

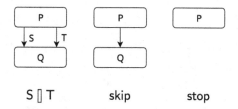

S [] T   skip   stop

3   State Diagrams for Concurrent Programming

Parallel Composition. The parallel composition $S_1 \parallel S_2$ is visualized by a dashed line:

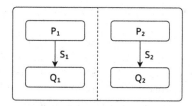

In the textual notation, *atomicity brackets* indicate how expressions are evaluated and statements executed as seen by concurrent programs. For example, $\langle x := x{+}1 \rangle$ means that x is incremented by 1 atomically, $\langle x \rangle := \langle x{+}1 \rangle$ means that first x is first read and then its new value written in separate steps. In state diagrams, each transition is executed atomically. Auxiliary variables ("registers") are used to express atomicity, for example:

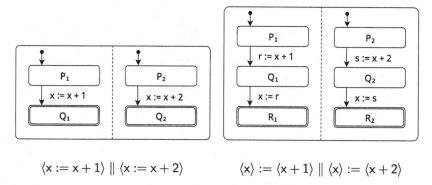

$\langle x := x + 1 \rangle \parallel \langle x := x + 2 \rangle$ $\langle x \rangle := \langle x + 1 \rangle \parallel \langle x \rangle := \langle x + 2 \rangle$

Non-interference. If S_1 and S_2 operate on distinct variables, the postcondition of $S_1 \parallel S_2$ is the conjunction of the postconditions of S_1 and S_2, as on the left-hand side below. If S_1 and S_2 operate on the same variables, as on the right-hand side below, it is not obvious what suitable pre- and postconditions of S_1 and S_2 would be:

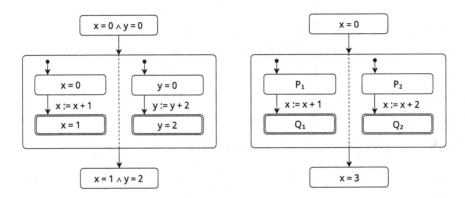

Annotation for x := x + 1 ∥ y := y + 2   Annotation for x := x + 1 ∥ x := x + 2

A naive approach, as on the left-hand side below, fails: if both processes start with x = 0 and x := x + 1 is executed first, the precondition of x := x + 2 is no longer x = 0. Likewise, if x := x + 2 is executed first and establishes x = 2, then x := x + 1 would invalidate that postcondition. The key is to weaken state predicates such that the execution in parallel processes will not invalidate the predicates, as to the right-hand side below:

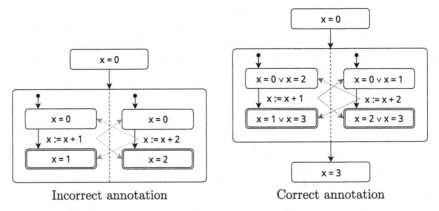

Incorrect annotation   Correct annotation

An atomic statement A that starts in state P *does not interfere* with predicate Q if A preserves Q, textually {P ∧ Q}A{Q}. For transitions, this is visualized as:

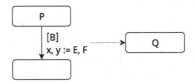

$$\text{if } P \wedge B \wedge Q \Rightarrow Q[x, y := E, F]$$

The rule for the *correctness of parallel composition* $S_1 \parallel S_2$ assumes that each process S_1, S_2 is correct in isolation:

$$\{P_1\}S_1\{Q_1\}$$
$$\{P_2\}S_2\{Q_2\}$$

Then the parallel composition will establish the conjunction of the postconditions under the conjunction of the preconditions,

$$\{P_1 \wedge P_2\}S_1 \parallel S_2\{Q_1 \wedge Q_2\}$$

provided that each atomic statement $S_{1,i}$ of S_1 preserves each state of S_2 and vice versa (the *non-interference condition*), visually:

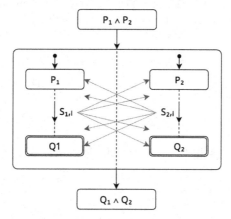

For example, the non-interference conditions for $x := x + 1 \parallel x := x + 2$ are derived from the green dotted lines of the correct annotation above:

1. $(x = 0 \vee x = 2) \wedge (x = 0 \vee x = 1) \Rightarrow (x = 0 \vee x = 1)[x := x + 1]$
2. $(x = 0 \vee x = 2) \wedge (x = 2 \vee x = 3) \Rightarrow (x = 2 \vee x = 3)[x := x + 1]$
3. ...

Condition Synchronization. The await statement "waits" until a Boolean expression is true and then executes its body atomically. The await statement is at the very core of every synchronization mechanism, even though programming languages do not support it in its full generality:

⟨await B then S⟩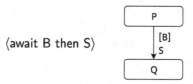

That is, an await is like an if with one alternative. For example, consider a producer who places objects into a (one-place) buffer. A consumer removes objects from the buffer. The producer has to wait until the buffer is empty before placing an object. Producers and consumers proceed in their own pace. The consumer has to wait until the buffer is not empty before removing the object:

$$\text{var } p, c : 0..N = 0, 0$$
$$\text{var } buf : T$$

var a : array 0..N − 1 of T = ...	var b : array 0..N − 1 of T
while p < N do	while c < N do
⟨await p = c⟩	⟨await p > c⟩
buf := a(p)	b(c) := buf
p := p + 1	c := c + 1

The state diagram below shows the required state annotations to allow to conclude that upon termination of Consumer, it will have a copy of array a of Producer in its own local array b.

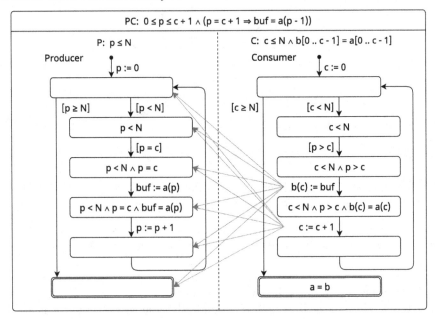

Each of the black solid transitions has to be correct in isolation. The producer invariant P does not contain variables that are modified by Consumer and likewise the consumer invariant C does not contain variables that are modified by Producer, so these are always preserved by transitions of the other process. The conditions for Consumer are, using correctness assertions:

1. $\{PC \wedge P \wedge C \wedge c < N \wedge p > c\} \; b(c) := buf \; \{PC \wedge C \wedge c < N \wedge p > c \wedge b(c) = a(c)\}$
2. $\{PC \wedge P \wedge C \wedge c < N \wedge p > c \wedge b(c) = a(c)\} \; c := c + 1 \; \{PC \wedge C\}$

The green dotted arrows indicate which transitions of Consumer may interfere with the states of Producer. There are six non-interference conditions for the transition with $b(c) := buf$; of those the three for the green dotted arrows going to "empty" states of Producer are identical, leaving four conditions:

1. $\{PC \wedge P \wedge C \wedge c < N \wedge p > c\}\ b(c) := buf\ \{true\}$
2. $\{PC \wedge P \wedge C \wedge c < N \wedge p > c \wedge p < N\}\ b(c) := buf\ \{p < N\}$
3. $\{PC \wedge P \wedge C \wedge c < N \wedge p > c \wedge p < N \wedge p = c\}\ b(c) := buf\ \{p < N \wedge p = c\}$
4. $\{PC \wedge P \wedge C \wedge c < N \wedge p > c \wedge p < N \wedge p = c \wedge buf = a(p)\}$
   $b(c) := buf$
   $\{p < N \wedge p = c \wedge buf = a(p)\}$

Likewise, there are four non-interference conditions for the transition with $c := c + 1$:

1. $\{PC \wedge P \wedge C \wedge c < N \wedge p > c\}\ c := c + 1\ \{true\}$
2. $\{PC \wedge P \wedge C \wedge c < N \wedge p > c \wedge p < N\}\ c := c + 1\ \{p < N\}$
3. $\{PC \wedge P \wedge C \wedge c < N \wedge p > c \wedge p < N \wedge p = c\}\ c := c + 1\ \{p < N \wedge p = c\}$
4. $\{PC \wedge P \wedge C \wedge c < N \wedge p > c \wedge p < N \wedge p = c \wedge buf = a(p)\}$
   $c := c + 1$
   $\{p < N \wedge p = c \wedge buf = a(p)\}$

The conditions for Producer not interfering with Consumer are analogous and have been left out for brevity.

Semaphores. A semaphore is an initialized integer variable with two operations, the wait operation P and the signal operation V, formally:

$$\text{var } s : \text{semaphore} = init$$
$$P(s) : \langle \text{await } s > 0 \text{ then } s := s - 1 \rangle$$
$$V(s) : \langle s := s + 1 \rangle$$

The *critical section* problem assumes that processes repeatedly try to enter a critical section, but only one is allowed to do. This can be enforced by using a binary semaphore, i.e. a semaphore whose value is either 0 or 1, for example for two concurrent processes:

$$\text{var mutex : semaphore} = 1$$

```
while true do          ||   while true do
  P(mutex)             ||     P(mutex)
  critical section     ||     critical section
  V(mutex)             ||     V(mutex)
  noncritical section  ||     noncritical section
```

To argue for the correctness, we add *ghost variables* in1, in2 to the two processes, CS1, CS2, that indicate if the process is in its critical section. Ghost variables are only assigned to and appear in invariants, but are not used in the program; obviously, they can be left out without affecting the program:

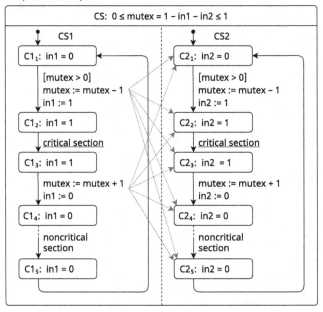

The *critical section property* is in1 + in2 ≤ 1, which is a consequence of CS above. The correctness of above program follows from each transition of CS1 being correct, i.e.

- $\mathsf{CS} \wedge \mathsf{C1}_1 \wedge \mathsf{mutex} > 0 \Rightarrow (\mathsf{CS} \wedge \mathsf{C1}_2)[\mathsf{mutex}, \mathsf{in1} := \mathsf{mutex} - 1, 1]$
- $\mathsf{CS} \wedge \mathsf{C1}_3 \Rightarrow (\mathsf{CS} \wedge \mathsf{C1}_4)[\mathsf{mutex}, \mathsf{in1} := \mathsf{mutex} + 1, 0]$

and each transition of CS1 not interfering with the states of CS2, as indicated by the green dotted arrows above, i.e.

- $\mathsf{CS} \wedge \mathsf{C1}_1 \wedge \mathsf{mutex} > 0 \wedge \mathsf{C2}_1 \Rightarrow \mathsf{C2}_1[\mathsf{mutex}, \mathsf{in1} := \mathsf{mutex} - 1, 1]$
- $\mathsf{CS} \wedge \mathsf{C1}_1 \wedge \mathsf{mutex} > 0 \wedge \mathsf{C2}_2 \Rightarrow \mathsf{C2}_2[\mathsf{mutex}, \mathsf{in1} := \mathsf{mutex} - 1, 1]$
- ...

The assumption is that neither <u>critical section</u> nor noncritical section contain operations on mutex. Because of symmetry, it then follows that each transition of CS2 is also correct and that CS2 does not interfere with the states of CS1.

4   Course Notes and Delivery

Jupyter notebooks, a web-based environment for interactive literate programming, are used for the course notes [14]. Notebooks consist of a sequence of *cells* with markdown or code. Markdown cells contain prose, algorithms, proofs, and diagrams, including state diagrams. A Jupyter extension was developed to ease formatting and colouring of algorithms and proofs, see Fig. 1. All diagrams,

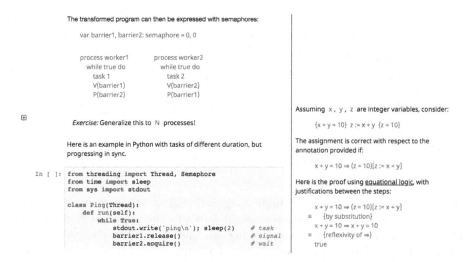

Fig. 1. (a) On the left a screenshot of a Jupyter notebook with an algorithm, an exercise whose solution can be revealed by clicking on the ⊞ symbol, and code cell with a Python implementation. (b) On the right a sequence of markdown cells.

including state diagrams, are created using a drawing editor. Currently, the code cells contain Python (for semaphores), Java (for monitors), and Go (for message passing) programs and can be executed right within the notebooks. The notebooks are available on a public GitHub repository[1] and on a local JupyterHub server. Students can either download the notebooks and run them on their own computers or run the notebooks on the JupterHub server in a web browser. JupyterHub supports grading of notebooks and is used for assignments and the final exam. JupyterHub also supports automated grading through test suites, however, because of the nondeterministic nature of concurrency, limited use is made of that. Still, the experience is that JupyterHub speeds up grading compared to the course management system that is used at McMaster, and is suitable for large classes.

The Jupyter notebooks are turned into slides by the RISE extension [4]; the course slides are the course notes with some cells suppressed. The slides retain the interactivity of Jupyter and allow code to be executed without leaving the slideshow. The experience is that the ability to switch quickly to executing code and back to explanations effectively keeps the attention of students even in larger classes.

5   Conclusions

Earlier on, the authors was using flowcharts to explain control structures, with annotations for (partial) correctness, following [21]. Despite the authors' best

---

[1] https://github.com/emilsekerinski/softwaredesign.

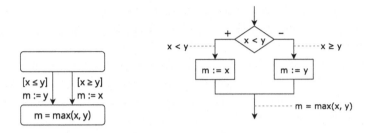

Fig. 2. State diagram vs flowchart

efforts, students produced too many spaghetti charts. The common flowchart notation cannot describe nondeterminism and blocking, see Fig. 2. Since the statements rather than the states are nested, there is no direct way to inherit invariants of enclosing states as with state diagrams, leading to repetition or extra naming of parts of annotations. By contrast, the overall experience with state diagrams for concurrent programming has been positive; state diagrams can give the formal rigour that concurrency calls for without being perceived as a "formal method" with its own language.

A tempting alternative is to use automated verification by model-checking, e.g. [6,8,15]. Another option is to use a static analysis tool for a language that is already used in the course to complement state diagrams. The author used ThreadSafe for Java programs in one year and this could be further explored [3].

References

1. Andrews, G.R.: Concurrent Programming: Principles and Practice. Benjamin/Cummings Publishing Company, Redwood City (1991)
2. Apt, K., de Boer, F.S., Olderog, E.-R.: Verification of Sequential and Concurrent Programs. Texts in Computer Science, 3rd edn. Springer, London (2009). https://doi.org/10.1007/978-1-84882-745-5
3. Atkey, R., Sannella, D.: ThreadSafe: static analysis for Java concurrency. Electron. Commun. EASST **72** (2015). https://doi.org/10.14279/tuj.eceasst.72.1025
4. Avila, D.: RISE: "Live" Reveal.js Jupyter/IPython Slideshow Extension: damianavila/RISE, June 2019. https://github.com/damianavila/RISE
5. Back, R.-J.: Invariant based programming. In: Donatelli, S., Thiagarajan, P.S. (eds.) ICATPN 2006. LNCS, vol. 4024, pp. 1–18. Springer, Heidelberg (2006). https://doi.org/10.1007/11767589_1
6. Ben-Ari, M.M.: A primer on model checking. ACM Inroads **1**(1), 40–47 (2010). https://doi.org/10.1145/1721933.1721950
7. Dijkstra, E.W., Feijen, W.H.J.: A Method of Programming. Addison-Wesley Publishing Company, Boston (1988)
8. Eisentraut, C., Hermanns, H.: pseuCo.com, June 2019. https://pseuco.com
9. Feijen, W., van Gasteren, A.J.M.: On a Method of Multiprogramming. Monographs in Computer Science. Springer, New York (1999). https://doi.org/10.1007/978-1-4757-3126-2

10. Gries, D., Schneider, F.B.: A Logical Approach to Discrete Math. Monographs in Computer Science. Springer, New York (1993). https://doi.org/10.1007/978-1-4757-3837-7

11. Harel, D.: Statecharts: a visual formalism for complex systems. Sci. Comput. Program. **8**, 231–274 (1987)

12. Kahl, W.: CALCCHECK: a proof checker for teaching the "logical approach to discrete math". In: Avigad, J., Mahboubi, A. (eds.) ITP 2018. LNCS, vol. 10895, pp. 324–341. Springer, Cham (2018). https://doi.org/10.1007/978-3-319-94821-8_19

13. Kernighan, B.W., Ritchie, D.M.: The C Programming Language. Prentice-Hall Inc., Upper Saddle River (1978)

14. Kluyver, T., et al.: Jupyter notebooks-a publishing format for reproducible computational workflows. In: Loizides, F., Schmidt, B. (eds.) Positioning and Power in Academic Publishing: Players, Agents and Agendas, pp. 87–90. IOS Press (2016). https://doi.org/10.3233/978-1-61499-649-1-87

15. Magee, J., Kramer, J.: Concurrency: State Models and Java Programs, 2nd edn. Wiley, New York (2006)

16. Manna, Z., Pnueli, A.: Temporal verification diagrams. In: Hagiya, M., Mitchell, J.C. (eds.) TACS 1994. LNCS, vol. 789, pp. 726–765. Springer, Heidelberg (1994). https://doi.org/10.1007/3-540-57887-0_123

17. Owicki, S., Gries, D.: An axiomatic proof technique for parallel programs. Acta Informatica **6**(4), 319–340 (1976). https://doi.org/10.1007/BF00268134

18. Ritchie, D.M.: The Development of the C language. In: The Second ACM SIGPLAN Conference on History of Programming Languages, HOPL-II, pp. 201–208. ACM, New York (1993). https://doi.org/10.1145/154766.155580

19. Sekerinski, E.: Design verification with state invariants. In: Lano, K. (ed.) UML 2 Semantics and Applications, pp. 317–347. Wiley, October 2009. https://doi.org/10.1002/9780470522622.ch13

20. Sekerinski, E.: Teaching the unifying mathematics of software design. In: Brouwer, R., Cukierman, D., Tsiknis, G. (eds.) Proceedings of the 14th Western Canadian Conference on Computing Education, WCCCE 2009, Burnaby, British Columbia, Canada, pp. 109–115. ACM, May 2009. https://doi.org/10.1145/1536274.1536307

21. Wirth, N.: Systematic Programming: An Introduction. Prentice-Hall, Englewood Cliffs (1973)

Teaching Discrete Mathematics
to Computer Science Students

Faron Moller$^{(\boxtimes)}$ and Liam O'Reilly

Swansea University, Swansea, UK
{F.G.Moller,L.P.OReilly}@swansea.ac.uk

Abstract. Discrete Mathematics is an inevitable part of any undergraduate computer science degree programme. However, computer science students typically find this to be at best a necessary evil with which they struggle to engage. Twenty years ago, we started to address this issue seriously in our university, and we have instituted a number of innovations throughout the years which have had a positive effect on engagement and, thus, attainment. At the turn of the century, a mere 2% of our first-year students attained a 1st-class mark (a mark over 70%) in the discrete mathematics course whilst over half of the class were awarded a failing grade (a mark under 40%). Despite the course syllabus and assessment remaining as difficult as ever (if not more challenging), and despite maintaining the same entrance requirements to the programme whilst more than tripling the class size, for the past two years, two-thirds of the class attained a first-class mark whilst less than 2% of the class failed. In this paper, we describe and motivate the innovations which we introduced, and provide a detailed analysis of how and why attainment levels varied over two decades as a direct result of these innovations.

1 Introduction

There are a great number of excellent textbooks for teaching computer science students the discrete mathematics which they will find necessary in their pursuit of the subject. Without prejudice, we can cite [6,9,17] as exemplars which have gone through multiple editions and commonly appear in the reading lists of relevant courses. However, whilst often written with computer science applications in mind, the standard presentation in such texts is inevitably mathematical in nature, with a methodical approach to formal syntax and semantics taking centre stage. As the modern computer science student often lacks the mathematical maturity of their predecessors (as argued below), this can be a hindrance to engagement and, thus, academic attainment.

That the modern computer science student is in general less mathematically minded than a generation ago is well recognised, and its causes now understood. Moller and Crick [13] give a detailed account of the history of computing education in UK schools: from a strong position in the 1980's with the introduction of the BBC Micro into every school along with a curriculum for teaching

© Springer Nature Switzerland AG 2019
B. Dongol et al. (Eds.): FMTea 2019, LNCS 11758, pp. 150–164, 2019.
https://doi.org/10.1007/978-3-030-32441-4_10

the fundamentals of programming including hardware, software, Boolean logic and number representation; through the 1990's and beyond where the emergence of pre-installed office productivity software led to the computing curricula being permeated – and overwritten – by basic IT skills; "Death-By-Powerpoint" became a common epithet for the subject. Beyond the arguments and references provided in [13], we can note a trend towards omitting mathematics as a prerequisite subject for studying computer science: of the 164 undergraduate computer science programmes offered by 105 universities in the UK, over 60% of these do not require mathematics as a school prerequisite [7].

There is a recognised digital skills shortage providing a high demand for computer science graduates [8], and an eagerness on the part of universities to fill places. However, with ever more students declaring in entrance statements that they are choosing to study computer science due to a love of digital devices rather than a love of the subject – and thus ever less prepared for the intellectual, logical and mathematical problem-solving challenges this entails – it can be a challenge in making some of the mathematical content of the curriculum palatable. This is especially true in the current climate where student satisfaction is a key indicator which universities are required by law in the UK to publish in their recruitment and marketing.

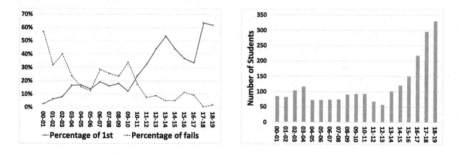

Fig. 1. Trends of students achieving 1st-class and failing results; and class sizes.

This paper describes an innovative approach that we have developed for teaching discrete mathematics to first-year university computer science students. By adopting and adapting our approach over the past twenty years from a traditional starting point, we have substantially increased the success rate – and substantially decreased the failure rate – of our students. Figure 1 shows how the percentage of students attaining a first-class mark (one over 70%) rose from 2% in 2000–2001 to over 60% in 2017–2018 and 2018–2019, whilst those failing the course (with a mark under 40%) dropped over the same time frame from 56% to under 2%. The figure also shows the class sizes which have more than tripled over the most recent five years which explains a noticeable dip in attainment which, we show, required further tweaking of our delivery model to address. The fact that this success is based on our approach is borne out by reflecting on annual student feedback for the various modules which students take across their programme of study; our delivery model is contrasted favourably against traditional approaches used in other modules taken by the same students, and recorded attendance (and hence engagement) is highest in this module.

2   Background

The nature of computer science education is changing, reflecting the increasing ubiquity and importance of its subject matter. In the last decades, computational methods and tools have revolutionised the sciences, engineering and technology. Computational concepts and techniques are starting to influence the way we think, reason and tackle problems; and computing systems have become an integral part of our professional, economic and social lives. The more we depend on these systems – particularly for safety-critical or economically-critical applications – the more we must ensure that they are safe, reliable and well designed, and the less forgiving we can be of failures, delays or inconveniences caused by the notorious "computer glitch."

Unlike for traditional engineering disciplines, the mathematical foundations underlying computer science are often not afforded the attention they deserve. The civil engineering student learns exactly how to define and analyse a mathematical model of the components of a bridge design so that it can be relied on not to fall down, and the aeronautical engineer learns exactly how to define and analyse a mathematical model of an aeroplane wing for the same purpose. However, software engineers are typically not as robustly drilled in the use of mathematical modelling tools. In the words of the eminent computer scientist Alan Kay [10], "most undergraduate degrees in computer science these days are basically Java vocational training." But computing systems can be at least as complex as bridges or aeroplanes, and a canon of mathematical methods for modelling computing systems is therefore very much needed. "Software's Chronic Crisis" was the title of a popular and widely-cited Scientific American article from 1994 [5] – with the dramatic term "software crisis" coined a quarter of a century earlier by Fritz Bauer [16] – and, unfortunately, its message remains valid a quarter of a century later.

University computer science departments face a sociological challenge posed by the fact that computers have become everyday, deceptively easy-to-use objects. Today's students – born directly into the heart of the computer era – have grown up with the Internet, a billion dollar computer games industry, and mobile phones with more computing power than the space shuttle. They often choose to study computer science on the basis of having a passion for using computing devices throughout their everyday lives, for everything from socialising with their friends to enjoying the latest films and music; and they often have less regard than they might to the considerations of what a university computer science programme entails, that it is far more than just *using* computers. In our experience, many of these students are easily turned off the subject when faced with a traditional course in discrete mathematics, with many of these, e.g., transferring into media or information studies. This has motivated us to reflect on our presentation of discrete mathematics, which has resulted in the following key considerations, all of which we have gleaned – and from which we have learned – from student feedback.

- *Do not rely on a service module provided by your mathematics department.* This is by no means a criticism of the mathematics department. It is simply the case that many students will not appreciate the importance of a course taken in a different department. At best, they may consider it peripheral to their studies, and at worst they will thus disengage completely.
- *Do not call it (discrete) mathematics.* A simple change of name from "discrete mathematics for computer science" to "modelling computing systems" in 2010–2011 was enough for us to witness a substantially increased level of engagement and attainment with the course, as made evident in Fig. 1. There was no other change that year to add to the cause of this effect.
- *Do not formalise early on.* The standard approach to, e.g., propositional logic is to present the formal syntax and semantics of the logic and emphasise the precise form and function of the connectives. The approach we have adopted is to stress the careful use of English, and to introduce logical symbols as mere shorthand for writing out English sentences. Formalism becomes far easier to adapt to if and once the students are comfortable with working with the concepts.
- *Exploit riddles and games.* As described later through characteristic examples, riddles and games provide an effective way to instil the rigours of computational thinking.
- *Use regular interactive small-group problem sessions.* We supplement three hours of weekly whole-class lectures with a one-hour small-group problem session (of 30–50 students) in which the emphasis is on the students carrying out computational problem-solving tasks, typically in pairs. We are confident in our thesis that this matters, as tweaking the sizes and regularity of these groups through the years coincides with peaks and dips in the attainment graphs. In particular, see the next consideration.
- *Keep these problem session groups small.* As can be seen in Fig. 1, attainment dropped between 2014 and 2017 as class sizes grew, but more than recovered in 2017–2018 despite a huge increase in the overall class size. This was due to an increase in the number of problem session groups; whilst the whole-class lectures became far less personable due to the huge numbers, the decrease in the sizes of the problem session groups resulted in much better results. Again, this being the only substantive change to delivery, we are confident in attributing the positive effect to this.

The first half of our course covers standard discrete mathematics topics: sets, propositional and predicate logics, functions and relations. Whilst it would be instructive to explore our approach to these topics, in this paper we explore our approach to teaching some of the topics from the latter part of the course. The reasons for this are two-fold. Firstly, the topics we discuss are typically not present in standard discrete mathematics courses; we make a case for why they ought to be so, for scientific reasons as well as due to the scope for presenting them in an engaging style. Secondly, our aim is to demonstrate the informal and engaging approach we take to the subject; we do so with the novel topics, leaving it to the readers' imagination as to how such techniques – e.g., the use of Smullyan- and Dudeney-style puzzles and riddles [4,18] – can be applied to the earlier standard topics.

3   Games and Winning Strategies

There is a long-standing tradition in disciplines like physics to teach modelling through little artefacts. The fundamental ideas of computational modelling and thinking as well can better be learned from idealised examples and exercises than from many real world computer applications. Our approach employs a large collection of logical puzzles and mathematical games that require no prior knowledge about computers and computing systems; these can be more fun and sometimes more challenging than, e.g., analysing a device driver or a criminal record database. Also, computational modelling and thinking is about much more than just computers.

In fact, games play a far more important role in our approach: they provide a novel approach to understanding computer software and systems. When a computer runs a program, for example, it is in a sense playing a game against the user who is providing the input to the program. The program represents a strategy which the computer is using in this game, and the computer wins the game if it correctly computes the result. In this game, the user is the adversary of the computer and is naturally trying to confound the computer, which itself is attempting to defend its claim that it is computing correctly, that is, that the program it is running is a winning strategy. (In software engineering, this game appears in the guise of *testing*.) Similarly, the controller of a software system that interacts with its environment plays a game against the environment: the controller tries to maintain the system's correctness properties, while the environment tries to confound them.

This view suggests an approach to addressing three basic problems in the design of computing systems:

1. *Specification* refers to the problem of precisely identifying the task to be solved, as well as what exactly constitutes a solution. This problem corresponds to the problem of defining a winning strategy.
2. *Implementation* or *synthesis* refers to the problem of devising a solution to the task which respects the specification. This problem corresponds to the problem of implementing a winning strategy.
3. *Verification* refers to the problem of demonstrating that the devised solution does indeed respect the specification. This problem corresponds to the problem of proving that a given strategy is in fact a winning strategy.

This analogy between the fundamental concepts in software engineering on the one hand, and games and strategies on the other, provides a mode of computational thinking which comes naturally to the human mind, and can be readily exploited to explain and understand software engineering concepts and their applications. It also motivates our thesis that game theory provides a paradigm for understanding the nature of computation.

4   Labelled Transition Systems

Labelled transition systems have always featured in the computer science curriculum, but traditionally (and increasingly historically) only in the context of finite automata within the study of formal languages. In our course we introduce them as general modelling devices, starting with an intuitively-clear and familiar use. Figure 2 presents Euclid's algorithm for computing the greatest common divisor of two numbers x and y, alongside a labelled transition system depicting the algorithm being hand-turned on the values 246 and 174.

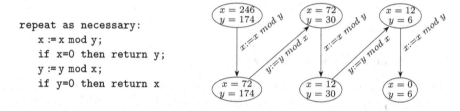

```
repeat as necessary:
    x := x mod y;
    if x=0 then return y;
    y := y mod x;
    if y=0 then return x
```

Fig. 2. Computing the greatest common divisor.

In general, a computation – or more generally a process – can be represented by a labelled transition system (LTS), which consists of a directed graph, where the vertices represent states, and the edges represent transitions from state to state, and are labelled by events. An LTS is typically presented pictorially as in Fig. 2, with the states represented by circles and the transitions by arrows between states labelled by actions.

As a further example, consider the lamp process depicted in Fig. 3. The lamp has a string to pull for turning the light on and off, and a reset button which

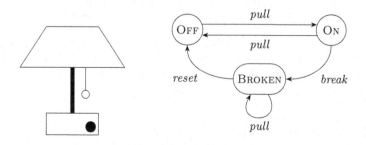

Fig. 3. The lamp process.

resets the circuit if a built-in circuit breaker breaks when the light is on. At any moment in time the lamp can be in one of three states:

- OFF – the light is off (and the circuit breaker is set);
- ON – the light is on (and the circuit breaker is set); and
- BROKEN – the circuit breaker is broken (and the light is off).

In any state the string can be pulled, causing a transition into the appropriate new state (from the state BROKEN, the new state is the same state BROKEN). In the state ON, the circuit breaker may break, causing a transition into the state BROKEN in which the reset button has popped out; from this state, the reset button may be pushed, causing a transition into the state OFF. (Note: discussions of design decisions naturally arise with the decision to always reset into the OFF state, regardless of the number of string pulls carried out in the BROKEN state. This provides a useful excursion into the problems that arise in the requirements analysis phase of software engineering.)

These two examples demonstrate the simple, but effective, use of LTSs as a means of modelling computing problems and real world objects.

4.1   Introducing LTSs with Puzzles

Whilst the definition of a labelled transition system is surprisingly straightforward for such a powerful formalism, getting students to engage with it requires some ingenuity. Fortunately, this is equally straightforward by resorting to well-known recreational puzzles.

4.2   The Man-Wolf-Goat-Cabbage Riddle

The following riddle was posed by Alcuin of York in the 8th century, and more recently tackled by Homer Simpson in a 2009 episode of The Simpsons titled Gone Maggie Gone.

> *A man needs to cross a river with a wolf, a goat and a cabbage. His boat is only large enough to carry himself and one of his three possessions, so he must transport these items one at a time. However, if he leaves the wolf and the goat together unattended, then the wolf will eat the goat; similarly, if he leaves the goat and the cabbage together unattended, then the goat will eat the cabbage. How can the man get across safely with his three items?*

The puzzle can be solved by modelling it as an LTS as depicted in Fig. 4. A state of the LTS will represent the current position (left or right bank) of the four entities (man, wolf, goat, cabbage); and there will be four actions representing the four possible actions that the man can take:

- m = the man crosses the river on his own;
- w = the man crosses the river with the wolf;
- g = the man crosses the river with the goat; and
- c = the man crosses the river with the cabbage.

The initial state is MWGC: (meaning all are on the left bank of the river), and we wish to find a sequence of actions which will lead to the state: MWGC (meaning all are on the right bank of the river). However, we want to avoid going through any of the six dangerous states WGC:M, GC:MW, WG:MC, MC:WG, MW:GC and M:WGC. There are several possibilities (all involving at least 7 crossings), for example: g, m, w, g, c, m, g.

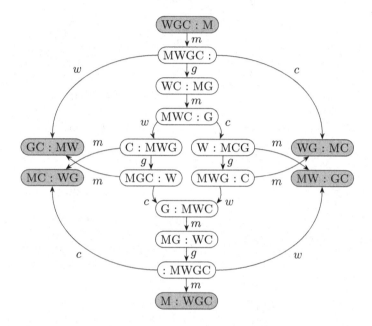

Fig. 4. The man-wolf-goat-cabbage LTS.

4.3   The Water Jugs Riddle

In the 1995 film Die Hard: With a Vengeance, New York detective John McClane (played by Bruce Willis) and Harlem dry cleaner Zeus Carver (played by Samuel L. Jackson) had to solve the following problem in order to prevent a bomb from exploding at a public fountain.

> *Given only a five-gallon jug and a three-gallon jug, neither with any markings on them, fill the larger jug with exactly four gallons of water from the fountain, and place it onto a scale in order to stop the bomb's timer and prevent disaster.*

This riddle – and many others like it – was posed by Abbot Albert in the 13th Century, and can be solved using an LTS. A state of the system underlying this riddle consists of a pair of integers (i, j) with $0 \leq i \leq 5$ and $0 \leq j \leq 3$, representing the volume of water in the 5-gallon and 3-gallon jugs A and B, respectively. The initial state is $(0,0)$ and the final state you wish to reach is $(4,0)$.

There are six moves possible from a given state (i,j) as listed in Fig. 5. Drawing out the LTS (admittedly a daunting task in this instance yet a useful exercise), we get the following 7-step solution:

$$(0,0) \xrightarrow{\textit{fillA}} (5,0) \xrightarrow{\textit{AtoB}} (2,3) \xrightarrow{\textit{emptyB}} (2,0) \xrightarrow{\textit{AtoB}} (0,2)$$
$$\xrightarrow{\textit{fillA}} (5,2) \xrightarrow{\textit{AtoB}} (4,3) \xrightarrow{\textit{emptyB}} (4,0).$$

$$(i,j) \xrightarrow{\textit{fillA}} (5,j) \qquad\qquad \textit{if} \quad i=0$$
$$(i,j) \xrightarrow{\textit{fillB}} (i,3) \qquad\qquad \textit{if} \quad j=0$$
$$(i,j) \xrightarrow{\textit{emptyA}} (0,j) \qquad\qquad \textit{if} \quad i>0$$
$$(i,j) \xrightarrow{\textit{emptyB}} (i,0) \qquad\qquad \textit{if} \quad j>0$$
$$(i,j) \xrightarrow{\textit{AtoB}} \big(\max(0,i+j-3),\min(3,i+j)\big) \qquad \textit{if} \quad i>0 \textit{ and } j<3$$
$$(i,j) \xrightarrow{\textit{BtoA}} \big(\min(5,i+j),\max(0,i+j-5)\big) \qquad \textit{if} \quad i<5 \textit{ and } j>0$$

Fig. 5. Water jug riddle moves.

These simple riddles and puzzles allow students to easily grasp and understand the powerful concept of labelled transition systems. After seeing only a few examples, they are able to model straightforward systems by themselves using LTSs. Once an intuitive understanding has been established, the task of understanding the mathematics behind LTSs becomes less foreboding.

5   Verification via Games

Having introduced a formalism for representing and simulating (the behaviour of) a system, the next question to explore is: *Is the system correct?* In its most basic form, this amounts to determining if the system matches its specification, where we assume that both the system and its specification are given as states of some LTS. For example, consider the two vending machines V_1 and V_2 depicted in Fig. 6, where V_1 is taken to represent the specification of the vending machine while V_2 is taken to represent its implementation. Clearly the behaviour of V_1 is somehow different from the behaviour of V_2: after *twice* inserting a 10p coin into V_1, we are *guaranteed* to be *able* to press the coffee button; this is *not* true of V_2. The question is: *How do we formally distinguish between processes?*

5.1   The Formal Definition of Equivalence

A traditional approach to this question relies on determining if these two states are related by a *bisimulation relation,* which is a binary relation R over its states in which whenever $(x, y) \in R$:

- if $x \xrightarrow{a} x'$ for some x' and a, then $y \xrightarrow{a} y'$ for some y' such that $(x', y') \in R$;
- if $y \xrightarrow{a} y'$ for some y' and a, then $x \xrightarrow{a} x'$ for some x' such that $(x', y') \in R$.

Simple inductive definitions already represent a major challenge for undergraduate university students; so it is no surprise that this coinductive definition of a bisimulation relation is incomprehensible even to some of the brightest postgraduate students – at least on their first encounter with it. It thus may seem incredulous to consider this to be a first-year discrete mathematics topic, even if it is a perfect application for exploring equivalence relations as taught earlier in the course. However, there is a straightforward way to explain the idea of bisimulation equivalence to first-year students – a way which they can readily grasp and are happy to explore and, indeed, play with. The approach is based on the following game.

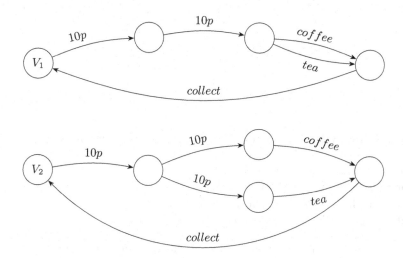

Fig. 6. Two vending machines.

5.2   The Copy-Cat Game

This game is played between two players, typically referred to as Alice and Bob. We start by placing tokens on two states of an LTS, and then proceed as follows.

1. Alice moves *either* of the two tokens forward along an arrow to another state; if this is impossible (that is, if there are no arrows leading out of either node on which the tokens sit), then Bob is declared to be the winner.

2. Bob must move the *other* token forward along an arrow which has *the same label* as the arrow used by Alice; if this is impossible, then the Alice is declared to be the winner.

This exchange of moves is repeated for as long as neither player gets stuck. If Bob ever gets stuck, then Alice is declared to be the winner; otherwise Bob is declared to be the winner (in particular, if the game goes on forever).

Alice, therefore, wants to show that the two states holding tokens are somehow different, in that there is something that can happen from one of the two states which cannot happen from the other. Bob, on the other hand, wants to show that the two states are the same: that whatever might happen from one of the two states can be copied by the other state.

It is easy to argue that two states should be considered equivalent exactly when Bob has a winning strategy in this game starting with the tokens on the two states in question; and indeed this is taken to be the definition of when two states are equal, specifically, when an implementation matches its specification.

As an example, consider playing the game on the LTS depicted in Fig. 7.

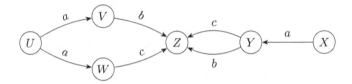

Fig. 7. A simple LTS.

Starting with tokens on states U and X, Alice has a winning strategy:

- Alice can make the move $U \xrightarrow{a} V$.
- Bob must respond with the move $X \xrightarrow{a} Y$.
- Alice can then make the move $Y \xrightarrow{c} Z$.
- Bob will be stuck, as there is no c-transition from V.

This example is a simplified version of the vending machine example; and a straightforward adaptation of the winning strategy for Alice will work in the game starting with the tokens on the vending machine states V_1 and V_2. We thus have an argument as to why the two vending machines are different.

5.3   Relating Winning Strategies to Equivalence

Whilst this notion of equality between states is particularly simple, and even entertaining to explore, it coincides precisely with the complicated coinductive definition of when two states are bisimulation equivalent. Seeing this is the case is almost equally straightforward.

- Suppose we play the copy-cat game starting with the tokens on two states x and y which are related by some bisimulation relation R. It is easy to see that Bob has a winning strategy: whatever move Alice makes, by the definition of a bisimulation relation, Bob will be able to copy this move in such a way that the two tokens will end up on states x' and y' which are again related by R; and Bob can keep repeating this for as long as the game lasts, meaning that he wins the game.
- Suppose now that R is the set of pairs of states of an LTS from which Bob has a winning strategy in the copy-cat game. It is easy to see that this is a bisimulation relation: suppose that $(x, y) \in R$:
  - if $x \xrightarrow{a} x'$ for some x' and a, then taking this to be a move by Alice in the copy-cat game, we let $y \xrightarrow{a} y'$ be a response by Bob using his winning strategy; this would mean that Bob still has a winning strategy from the resulting pair of states, that is $(x', y') \in R$;
  - if $y \xrightarrow{a} y'$ for some y' and a, then taking this to be a move by Alice in the copy-cat game, we let $x \xrightarrow{a} x'$ be a response by Bob using his winning strategy; this would mean that Bob still has a winning strategy from the resulting pair of states, that is $(x', y') \in R$.

We have thus taken a concept which baffles postgraduate research students, and presented it in a way which is well within the grasp of first-year undergraduate students.

5.4   Determining Who Has the Winning Strategy

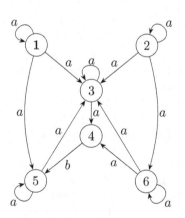

Once the notion of equivalence is understood in terms of winning strategies in the copy-cat game, the question then arises as to how to determine if two particular states are equivalent, ie, if Bob has a winning strategy starting with the tokens on the two given states. This isn't generally a simple prospect; games like chess and go are notoriously difficult to play perfectly, as you can only look ahead a few moves before getting caught up in the vast number of positions into which the game may evolve.

Here again, though, we have a straightforward way to determine when two states are equivalent. Suppose we could paint the states of an LTS in such a way that any two states which are equivalent – that is, from which Bob has a winning strategy – are painted the same colour. The following property would then hold.

If any state with some colour C has a transition leading out of it into a state with some colour C', then every state with colour C has an identically-labelled transition leading out of it into a state coloured C'.

That is, if two tokens are on like-coloured states (meaning that Bob has a winning strategy) then no matter what move Alice makes, Bob can respond in such a way as to keep the tokens on like-coloured states (ie, a position from which he still has a winning strategy). We refer to such a special colouring of the states a *game colouring*.

To demonstrate, consider the following LTS.

At the moment, all states are coloured white, and we might consider whether this is a valid game colouring. It becomes readily apparent that it is not, as the white state 4 can make a b-transition to the white state 5 whereas none of the other white states (1, 2, 3, 5 and 6) can do likewise. In fact, in any game colouring, the state 4 must have a different colour from 1, 2, 3, 5 and 6.

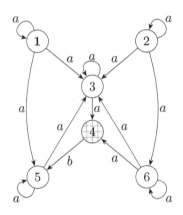

Hence we paint it a different colour from white; in order to present this example in black-and-white, we shall paint the state 4 with the new colour "checkered."

(Of course, unbound in class by black-and-white printing, we'd use an actual colour in practice). We again consider whether this is now a valid game colouring. Again it becomes apparent that it is not, as the white states 3 and 6 have a-transitions to a checkered state, whereas none of the other white states 1, 2 and 5 do. And in any game colouring, the states 3 and 6 must have a different colour from 1, 2 and 5.

Hence we paint these a different colour from white and checkered; we shall choose the colour "swirly."

We again consider whether this is now a valid game colouring. This time we find that it is, as every state can do exactly the same thing as every other state of the same colour: every white state has an a-transition to a white state and an a-transition to a swirly state; every swirly state has an a-transition to a swirly state and an a-transition to a checkered state; and every checkered state has a b-transition to a white state.

At this point we have a complete understanding of the game, and can say with certainty which states are equivalent to each other. This is an exercise which first-year students can happily carry out on arbitrarily-complicated LTSs, which again gives testament to the effectiveness of using games to great success in imparting difficult theoretical concepts to first-year students – in this case the concept of partition refinement.

6   Conclusion

We teach first-year Discrete Mathematics in the guise of modelling computing systems; and we find that our students quickly and easily understand the modelling of computing systems when it is done in a way which nurtures their willingness to engage. Starting with formal syntax and semantics and complicated real world examples, in our experience, makes the task very daunting, difficult and generally unpleasant for students. However, appealing to their existing understand of how the world works, using puzzles as a medium, students can quickly become comfortable using mathematical concepts such as LTSs. A similar lesson is learnt when it comes to teaching verification: starting with the formal definition of bisimulation (or similar) is an uphill battle from the start, even for postgraduate research students. However, starting from games like the copy-cat game, such topics become immediately accessible.

We have used this approach for over a decade to teach discrete mathematics incorporating the modelling and verification of computing systems as part of our first-year undergraduate programme, resulting in the publication of the course textbook [14]. With the fine-tuning of our approach, and abiding by the considerations outlined in Sect. 2, we have succeeded in maximising attainment levels of the students through active and interested engagement.

Of course, problem solving through recreational mathematics – which is ultimately what we are exploiting in our approach – has very many proponents, and there is a long and extensive history of books marketed towards the mathematically-inquisitive. We are by no means alone in recognising the power of applying recreational mathematics to the development of computational problem solving skills; as relevant exemplars we note Averbach and Chein's *Problem Solving Through Recreational Mathematics* [1], Backhouse's *Algorithmic Problem Solving* [2], Levitin and Levitin's *Algorithmic Puzzles* [11]; and Michalewicz and Michalewicz's *Puzzle-Based Learning* [12]. What we offer in particular is an embedding of the approach from day one of the first year of our students' undergraduate journey, in particular to engage them in a topic – discrete mathematics – that they typically struggle with, both academically and in terms of recognising its relevance in the subject. In this sense, we are closely related to the various approaches that have been developed of late for introducing school-aged audiences to computational thinking. In this vein we note the CS Unplugged[1] and the CS4Fun[2] initiatives. Indeed, much of our material has been adapted into school workshops for the Technocamps[3] initiative.

The "informal" way in which we approach the teaching of formal methods has many parallels with Morgan's *(In)Formal Methods: The Lost Art* [15]. The course described in this report is for upper-level computer science students who are already adept at writing programs who are studying software development methods, whereas our course is for first-year students and thus very

---

[1] csunplugged.org.

[2] cs4fn.org.

[3] technocamps.com.

much preliminary. Nonetheless, many of the findings in [15] – in particular as reflected in the student feedback – are replicated in our course, where positive feedback is provided on: the interactive and hands-on approach; the amusing exercises and assignments; the class room style teaching; the overall teaching methodology with dedicated tutors; and the means by which the relevance of the course is stressed.

As a final note, many of the considerations that we have identified as being important in teaching mathematics to computing students are reflected in [3] as being useful and thus adopted in their novel approach to teaching computing to mathematics students.

References

1. Averbach, B., Chein, O.: Problem Solving Through Recreational Mathematics. Dover, Mineola (1980)
2. Backhouse, R.: Algorithmic Problem Solving. Wiley, Hoboken (2011)
3. Betteridge, J., et al.: Teaching of computing to mathematics students. In: Proceedings of the 3rd Conference on Computing Education Practice, CEP 2019, Durham, UK, 9 January 2019, pp. 12:1–12:4 (2019)
4. Dudeney, H.: The Canterbury Puzzles, and Other Curious Problems. Andesite Press (2015)
5. Gibbs, W.W.: Software's chronic crisis. Sci. Am. 271(3), 86–95 (2004)
6. Hein, J.L.: Discrete Structures, Logic, and Computability, 4th edn. Jones and Bartlett, Burlington (2017)
7. Higher Education Statistics Agency (HESA): Recruitment data for computer science courses in the UK (2019). https://www.hesa.ac.uk
8. House of Commons Science and Technology Committee: Digital Skills Crisis: Second Report of Session 2016–2017, 7 June 2016
9. Johnsonbaugh, R.: Discrete Mathematics, 8th edn. Pearson, New York (2017)
10. Kay, A.: A conversation with Alan Kay. ACM Queue 2(9), 20–30 (2004)
11. Levitin, A., Levitin, M.: Algorithmic Puzzles. Oxford University Press, Oxford (2011)
12. Michalewicz, Z., Michalewicz, M.: Puzzle-Based Learning: An Introduction to Critical Thinking, Mathematics, and Problem Solving. Hybrid Publishers, Melbourne (2010)
13. Moller, F., Crick, T.: A university-based model for supporting computer science curriculum reform. J. Comput. Educ. 5(4), 415–434 (2018)
14. Moller, F., Struth, G.: Modelling Computing Systems Mathematics for Computer Science. Springer, London (2013). https://doi.org/10.1007/978-1-84800-322-4
15. Morgan, C.: (In-)formal methods: the lost art. In: Liu, Z., Zhang, Z. (eds.) SETSS 2014. LNCS, vol. 9506, pp. 1–79. Springer, Cham (2016). https://doi.org/10.1007/978-3-319-29628-9_1
16. Naur, P., Randell, B. (eds.): Software Engineering: Report of a conference sponsored by the NATO Science Committee, Garmisch, Germany, 7–11 October 1968. NATO Scientific Affairs Division (1969)
17. Rosen, K.H.: Discrete Mathematics and Its Applications, 8th edn. McGraw-Hill, New York (2019)
18. Smullyan, R.: To Mock a Mockingbird: And Other Logic Puzzles. Oxford University Press, Oxford (2000)

Principled and Pragmatic Specification of Programming Languages

Adrian Johnstone$^{(\boxtimes)}$ [ID] and Elizabeth Scott [ID]

Royal Holloway, University of London, Egham, Surrey TW20 0EX, UK
{a.johnstone,e.scott}@rhul.ac.uk

Abstract. Programmers from the imperative tradition often have little experience of using inductive definitions and inference, and that may explain why executable SOS specifications have not become a standard feature of mainstream language development toolkits. We wish to 'de-mystify' SOS for such programmers, allowing precise and principled specifications to be given for even small industrial DSL's. eSOS (elided Structural Operational Semantics) is a compact tool for specifying executable formal semantics. It is designed to be a translation target for enriched SOS specification languages. The simplicity of eSOS and its reference Java implementation allow programmers to follow the details of an execution trace, and to step through rules using a conventional debugging framework, allowing them to understand and use SOS-based specifications to construct usable language interpreters.

Keywords: Structural Operational Semantics · Domain Specific Language specification · Operationalising formal specifications

1 Introduction

Formal specification of programming language semantics is still seen by most software engineers as an esoteric and opaque approach to language implementation. In this paper we describe our approach to 'de-mystifying' formal semantics by embedding a simple model of SOS interpretation into a final year course on the engineering of Domain Specific Languages. We use the formal specification of semantics as a concise and precise specification from which an interpreter may be automatically generated rather than emphasising verification or proving properties of programs. We limit ourselves to sequential languages and as a result direct interpretation of the rules can yield processors which are fast enough for many applications.

A key part of the approach is to show how to write the SOS rule interpreter itself in a few lines of a procedural programming language. This allows practitioner programmers who may have little or no mathematical training to understand operationally how the specification is executed, and to think of formal specification as 'just another kind of programming'. We have found that the approach is successful even with 'maths-averse' students.

© Springer Nature Switzerland AG 2019
B. Dongol et al. (Eds.): FMTea 2019, LNCS 11758, pp. 165–180, 2019.
https://doi.org/10.1007/978-3-030-32441-4_11

Is this sort of programmer-driven approach necessary or even desirable? Well, it is nearly 40 years since Plotkin introduced Structural Operational Semantics (SOS) in a series of lectures at Aarhus University [6]. In that time the computer science theory community has generated myriad related papers, and SOS is now firmly established as part of the basic toolkit for any programming language researcher. Perhaps surprisingly, the practitioner community has in large measure eschewed formal semantics, including SOS. This is in marked contrast to, say, syntax definition using BNF, formal parsing algorithms and even attribute grammars which are widely used (at least in the somewhat informal way that Bison and some other parsing toolkits provide). This ought not to be the case: the core ideas of SOS are certainly no harder to grasp than the notions of grammars and derivation trees.

We would consider that SOS had entered the mainstream if some of the following were true: textbooks on languages and compilers had a chapter on using SOS to write precise descriptions of some or all of a language; widely used language implementation toolkits included a SOS specification capability and an associated interpreter; programming language standards used at least a little SOS to clarify details; and online forums featured discussion of the pragmatics of applying SOS. In fact none of these are true – for instance on the Stack Overflow forums there are a few questions about SOS and its place in the spectrum of formal semantics techniques, but almost none of the pragmatic 'how do I do X in Y' questions that characterise the forums for widely deployed software tools.

Why might this be? It could be that SOS specifications quickly become too large to be useful, but current informal programming language standards documents are hardly noted for brevity. We suspect that the problem simply arises from the usual cultural gap in our discipline: that in practice the entry price for understanding declarative specifications comprising inductive definitions of relations via inference rules is too high for many working procedural programmers.

Our hypothesis, then, is that if we could reduce SOS interpretation to a simple procedural operation over the rules, programmers would embrace the brevity and clarity of the approach. The problem seems to be that the core idea of inference and the heavy use of mathematical notation in typical SOS textbooks are offputting to 'normal' programmers: perhaps ironically, they simply don't understand the meaning of the semantic formalism.

Several ambitious projects aim to deliver the benefits of formal semantics specification in a programmer-friendly manner; including the well known and now-venerable ASF+SDF system [10] (and its successors including RascalMPL [4] and the Spoofax [3] language workbench), the K system [7] and tools such as OTT [9]. The PlanCompS project [1] provides a unifying approach by abstracting away from formal semantics frameworks, building specifications from small *fundamental constructs*.

Our eSOS system, described in this paper, certainly does not compete with these rich systems, but rather attempts to leverage procedural programmers' existing knowledge to give them a way into formal semantics. The core tool comes as two small Java packages, one containing a value system which provides straightforward runtime type checking, and the other containing a parser for eSOS specifications, classes implementing an abstract syntax for SOS, and an SOS interpreter for sequential programs. Though conceived as a back end for richer SOS notations, we have found that it is a comfortable notation for neophyte users who can think of it as just another form of programming, and can answer questions such as 'yes, but what does that really *do*' by looking at the source code and exercising SOS specifications using the debugger from their preferred development environment to walk through traces.

2   The Course

We run a third year course entitled *Software Language Engineering (SLE)*. This was originally conceived as a pragmatic counterpart to our existing compiler theory course which presents topics in parsing and code generation and optimisation. The SLE course is intended to equip students with the engineering skills needed to design and deliver a fully working interpreter or compiler for a small language, with no emphasis on optimisation: the primary goal is correctness of the language processor, not high performance. The focus is on Domain Specific Languages, with motivating examples which include 3D modelling languages for graphics and 3D printing; music specification languages which connect to the Java MIDI synthesizer; and image processing languages. As part of the course, students develop their own DSL's, usually in one of these domains.

Over time, the two courses have developed independently, and now not all of the SLE students take the compiler theory course. As a result, we work *ab initio* and make no assumptions about prior knowledge of compilers.

The students are rarely mathematically confident. They will have taken typical first year courses on discrete maths with some exposure to logic and the use of inference rules, but they will not have previously applied that knowledge beyond very small pencil-and-paper exercises.

The course is taught over ten weeks (plus one week of revision and consolidation) each of which has two one-hour lectures and a two-hour lab session. There are seven programmed labs; the remaining sessions are used for tutorial support whilst the students develop their own language projects.

The first week is critical. The goal is to de-mystify formal systems by presenting rule based 'symbol-pushing' games. We use Conway's Game of Life as an example. Our students are familiar with this formal system because the first large program that they write in year one is a graphical version of Life.

We then need to help students become comfortable with a reduction model of program execution in which the program is progressively rewritten (with side-effects recorded in semantic *entities* such as the store and the environment). Most students have a von Neumann mind set in which a static program is traversed under the control of the program counter: we tell them that we need to 'get rid of the program counter' before we can use our chosen formal specification method.

We introduce the idea of establishing a program's meaning by repeatedly rewriting it, rather than by an execution walkthrough, using a version of Euclid's

Greatest Common Divisor (GCD) algorithm written in a simple procedural language with implicit declarations:

```
a := 15; b := 9;
while a != b
  if a > b
    a := a - b;
  else
    b := b - a;
gcd := a;
```

The program leaves its result in variable gcd; with a and b initialised to 15 and 9 we expect that after execution gcd would contain 3.

We also give the program in an internal (abstract) syntax form, as a term built from prefix functions. The internal abstract syntax is not formally defined at this stage: we simply use a form that is sufficiently close to the concrete program that students can accept it as being equivalent.

```
seq(seq(seq(
  assign(a, 15), assign(b, 9)),
  while(ne(deref(a), deref(b)),
    if(gt(deref(a), deref(b)),
      assign(a, sub(deref(a), deref(b))),
      assign(b, sub(deref(b), deref(a)))))),
  assign(gcd, deref(a)))
```

We then compare the behaviour of the concrete program (as observed via a walkthrough in the Eclipse debugger) with the behaviour of the internal abstract syntax term under term rewriting. This is a purely illustrative exercise, but nevertheless sufficient to informally show that term rewriting can mimic the execution of the program as conventionally understood.

In the main body of the course, students learn four key techniques: eSOS interpretation; parsing; attribute grammar evaluation; and (limited) term rewriting. In each case, the technique is presented as a formal system, but with an accompanying procedural model rendered in Java code. Often the procedural model presented in lectures is not fully general but is sufficient to provide an intellectual model that allows them to use more powerful versions of the same idea in the labs as a black box, without being burdened by their internal complexity.

Parsing forms a good example of this style of *learn-by-doing, and take-the-rest-on-trust*. For project work, the production parsing technology that we use is the GLL generalised parser [8] but a detailed description of that method would require too much classroom time. Instead, students learn how to hand-write in Java simple singleton-backtracking recursive descent parsers. We then look at grammatical constructs for which that approach fails: we believe that things which are broken can be more educational than things that seem to magically work. The students go on to use a GLL parser which behaves to some extent like a backtracking parser but overcomes these problems in a way they don't need to know the details of. We extend the parsers to support attributes and attribute equations, and use the resulting tool to implement a grammar for BNF itself with attribute equations which generate. By the end of this two week segment the students have developed a bootstrapped parser generator which can reproduce itself. They understand parsing, meta-description and generation of programs from specifications. Students know that the parsing technology they have explored is weak, but understand that the principles scale up directly to more general parsing approaches.

3 How We Teach SOS

We teach SOS using eSOS, a variant we have developed to be accessible to mainstream students with a basic procedural programming background. In the later sections we shall describe the eSOS interpreter that allows students to experiment by executing their specifications. However, we don't just want to talk about teaching, we want to illustrate our classroom style. The SLE course has an accompanying textbook which is being developed as we gain experience and feedback from the course. In this section we provide a precis of the lecture material which introduces eSOS to illustrate the way in which we strip the subtleties down to a basic minimum. The rest of Sect. 3 is written as though for students. This allows the reader who is not a SOS practitioner to pick up the notions and terminology they need for the subsequent sections. Experts will find nothing surprising and can skip to the description of the eSOS interpreter in Sect. 4.

At hardware level, computer programs exist as essentially static patterns of instructions, traversed under the control of a *program counter* which forms a pointer into the program. It can be difficult to directly prove properties of programs in this model, since the evolution of the computation is a property of the trace of the program counter. An initial step in formalising programming language semantics is often to move to a 'reduction' model, in which the program is a dynamic object that may be rewritten during execution. Most (though not all) execution steps reduce the size of the program term. For pure functional programming languages, these rewrites capture everything there is to say about the computation, but most languages also allow side-effects such as store updates and appends to output lists.

Here is the four step reduction of a program term which computes $10 - 2 - 4$ and 'outputs' the result by appending it to an initially empty list.

```
<output(sub(sub(10, 2),4)), []>
<output(sub(8,4)), []>
<output(4), []>
<, [4]>
```

At each point, some part of the program term called the *reducible expression* or redex has been identified, a simple computation performed and then the term rewritten: in the first step the redex sub(10, 2) has been rewritten to 8.

A configuration ⟨program term, output list⟩ thus captures everything about the state of the computation at some point: the list captures side-effects of previous computations and the program term contains what is left to be computed. The output list is an example of a *semantic entity*: depending on the style of language we are specifying, configurations may have several entities in addition to the program term.

In eSOS we have five kinds of entity: (i) read-only lists and (ii) write-only lists model input and output; (iii) maps whose bindings may be changed which model read/write memory (usually called *stores*); (iv) maps whose bindings may not changed which model symbol tables (or *environments*) and (v) singleton sets which are used for describing signals and exceptions. We refer to a ⟨program term, entity list⟩ pair as a *configuration*. If the entity list is empty, we may omit it.

Execution of programs, then, is modeled by stepping from configuration to configuration. The components of a configuration vary according to the language being specified. In the subtraction example we have a program term and a write-only list. Our abstract internal form of the GCD program above does not have input and output, so all we need is a program term and a store, denoted as $\langle \theta, \sigma \rangle$.

We can view program execution as a sequence of configuration transitions: configuration X transitions to Y if there is a program whose transition sequence has Y appearing as a successor to X. The set of all transitions describes everything that could possibly be executed: in a deep sense it *is* the semantics of the language of those programs.

An SOS specification is merely a device for specifying a (usually infinite) set of transitions using a finite recipe of *inference rules*. For languages with configurations $\langle \theta, \sigma \rangle$, each rule has the form

$$\frac{C_1 \quad C_2 \quad \ldots \quad C_k}{\langle \theta, \sigma \rangle \rightarrow \langle \theta', \sigma' \rangle}$$

The single transition below the line is the *conclusion*. The C_i are the *conditions*: there may be zero or more of them. Conditions can themselves be transitions, or may be functions. The latter are referred to as *side-conditions*.

One might read an inference rule in this style as:

if you have a configuration $\langle \theta, \sigma \rangle$,
and C_1 succeeds and C_2 succeeds and ... and C_k succeeds
then we can transition to configuration $\langle \theta', \sigma' \rangle$

One uses this kind of rule by checking that the current configuration matches the left hand side of the conclusion, then checking the conditions (in any order) and then, if everything succeeds, rewriting the current configuration into the right hand side of the conclusion. Where a condition is itself a transition we must recursively apply our checking process to transitions in the conditions. The subchecking can only terminate when we encounter a rule with no transitions in its conditions.

In practice, to produce a finite specification, SOS rules are written as *rule schemas* in which variables are used as placeholders for subterms. For example,

$$<\mathsf{seq}(\mathit{done}, C)> \rightarrow <C>$$

is a rule schema with variable C, and a rule is obtained by replacing C with a program term

$$<\mathsf{seq}(\mathit{done}, \mathsf{output}(6))> \rightarrow <\mathsf{output}(6)> \,.$$

When interpreting these rule schemas, we use the operations of *pattern matching* and *substitution* to dissect and reconstruct terms. We call a term which contains variables an *open term* or *pattern*. A term with no variables is *closed*.

We shall write $\theta \triangleright \pi$ for the operation of matching closed term θ against pattern π. The result of such a pattern match is either *failure* represented by $\bot$, or a set of bindings. So, in these expressions where X is a variable

$$\mathsf{seq}(\mathsf{done},\ \mathsf{output}(6)) \triangleright \mathsf{seq}(\mathsf{done},\ X)$$

returns $\{X \mapsto \mathsf{output}(6)\}$ whereas

$$\mathsf{seq}(\mathsf{done},\ \mathsf{output}(6)) \triangleright \mathsf{seq}(X,\ \mathsf{done})$$

returns $\bot$ because $\mathsf{output}(6)$ does not match done.

Pattern substitution is the process of substituting bound subterms for the variables in the pattern. We shall write $\pi \triangleleft \rho$ for the operation of replacing variables in pattern π with their bound terms from ρ. So

$$\mathsf{plus}(X,\ 10) \triangleleft \{X \mapsto 6\} \ \text{ returns } \mathsf{plus}(6,\ 10).$$

The following SOS rule (schema) handles the subtraction of two integers. It has three side conditions which use pre-specified functions isInt and subOp. The construct sub belongs to the abstract syntax of the language whose semantics are being specified.

$$\frac{\mathtt{isInt}(n_1) \triangleright \mathit{true} \ \ \mathtt{isInt}(n_2) \triangleright \mathit{true} \ \ \mathtt{subOp}(n_1, n_2) \triangleright V}{\langle \mathsf{sub}(n_1, n_2) \rangle \to \langle V \rangle} \qquad [\text{sub}]$$

The conclusion tells us that this rule will rewrite expressions of the form sub(n1, n2) to some value, if the conditions (which are all side-conditions) are met.

How should we use such rules to implement interpreters? Let us assume that the current program term is θ, then one way to compute whether the transition may be made is:

> **if** $\rho_1 = (\theta \triangleright \mathsf{sub}(n_1, n_2))$ **then**
> **if** $(\mathtt{isInt}(n_1) \triangleleft \rho_1) \triangleright \mathit{true}$
> **and** $(\mathtt{isInt}(n_2) \triangleleft \rho_1) \triangleright \mathit{true}$
> **and** $\rho_2 = ((\mathtt{subOp}(n_1, n_2) \triangleleft \rho_1) \triangleright V)$
> **then** $\theta' = V \triangleleft \rho_2$

Informally, we try to match the current program term against the left-hand side of the conclusion and store any variable bindings in the map ρ_1. We then work through the conditions substituting for variables on their left hand sides and perhaps creating new environments for pattern matches. If all of the conditions succeed then we make a new program term θ' by substituting the most recent environment. If we can guarantee that each variable appears only once as an argument to a pattern match operator $\triangleright$, then we can use a single environment which is extended as we work through the conditions.

Using the two rules below, we specify program terms which are nested subtractions.

$$\frac{\langle E_1, \alpha \rangle \rightarrow \langle I_1, \alpha \rangle}{\langle \mathsf{sub}(E_1, E_2), \alpha \rangle \rightarrow \langle \mathsf{sub}(I_1, E_2), \alpha \rangle} \qquad \text{[subLeft]}$$

$$\frac{\langle E_2, \alpha \rangle \rightarrow \langle I_2, \alpha \rangle \quad \mathsf{isInt}(n) \triangleright \mathit{true}}{\langle \mathsf{sub}(n, E_2), \alpha \rangle \rightarrow \langle \mathsf{sub}(n, I_2), \alpha \rangle} \qquad \text{[subRight]}$$

The rule [subLeft] rewrites the left argument to a simpler expression whilst preserving the second argument. Rule [subRight] will only process terms that had a single integer as the left hand argument, and rewrites the second argument. The original [sub] rule will then perform the subtraction of the integers. Together these three rules comprise a so-called *small-step* SOS for subtraction and act so as to enforce left to right parameter evaluation order.

When running the interpreter on a particular initial term, we can put in checks to ensure that at most one rule is activated at each rewrite step, though of course that will only detect non-determinism that is triggered by that particular term. Static checking of rules can detect some forms of non-determinism.

Example: SOS Rules for the GCD Internal Language

An SOS specification may name more than one set of transitions. The rules we have looked at so far are so-called 'small-step' rules. Big-step rules in which, say, arithmetic operations proceed directly to their result without the fine-grained elaboration of the left and right arguments are also possible, and both types of transition may occur within one set of rules. We illustrate this technique with a complete set of rules for our GCD abstract internal language in which the relational and arithmetic operations are specified using a big-step transition $\Rightarrow$ and the commands using a small-step transition $\rightarrow$. It is sometimes helpful to think of small-step rules such as [assignResolve] 'calling' the big step transition to reduce a complex arithmetic expression to a value.

As well as arithmetic and boolean values, this specification uses the special value done (sometimes called skip in the literature) which represents the final reduction state of a program term.

$$< \mathsf{seq}(\mathit{done}, C), \sigma > \rightarrow < C, \sigma > \qquad \text{[sequenceDone]}$$

$$\frac{< C_1, \sigma > \to < C_1', \sigma' >}{< \mathsf{seq}(C_1, C_2), \sigma > \to < \mathsf{seq}(C_1', C_2), \sigma' >} \quad \text{[sequence]}$$

$$< \mathsf{if}(\mathit{true}, C_1, C_2), \sigma > \to < C_1, \sigma > \quad \text{[ifTrue]}$$

$$< \mathsf{if}(\mathit{false}, C_1, C_2), \sigma > \to < C_2, \sigma > \quad \text{[ifFalse]}$$

$$\frac{< E, \sigma > \Rightarrow < E', \sigma' >}{< \mathsf{if}(E, C_1, C_2), \sigma > \to < \mathsf{if}(E', C_1, C_2), \sigma' >} \quad \text{[ifResolve]}$$

$$\frac{< \mathsf{if}(E, \mathsf{seq}(C, \mathsf{while}(E, C)), \mathit{done}), \sigma > \to < C', \sigma' >}{< \mathsf{while}(E, C), \sigma > \to < C', \sigma' >} \quad \text{[while]}$$

$$\frac{\mathtt{isInt}(n) \triangleright \mathit{true} \quad \mathtt{updateOp}(\sigma, X, n) \triangleright \sigma_1}{< \mathsf{assign}(X, n), \sigma > \to < \mathit{done}, \sigma_1 >} \quad \text{[assign]}$$

$$\frac{< E, \sigma > \Rightarrow < n, \sigma' >}{< \mathsf{assign}(X, E), \sigma > \to < \mathsf{assign}(X, n), \sigma' >} \quad \text{[assignResolve]}$$

$$\frac{< E_1, \sigma > \Rightarrow < n_1, \sigma_1 > \; < E_2, \sigma_1 > \Rightarrow < n_2, \sigma_2 > \; \mathtt{gtOp}(n_1, n_2) \triangleright V}{< \mathsf{gt}(E_1, E_2), \sigma > \Rightarrow < V, \sigma_2 >} \quad \text{[gtBig]}$$

$$\frac{< E_1, \sigma > \Rightarrow < n_1, \sigma_1 > \; < E_2, \sigma_1 > \Rightarrow < n_2, \sigma_2 > \; \mathtt{neOp}(n_1, n_2) \triangleright V}{< \mathsf{ne}(E_1, E_2), \sigma > \Rightarrow < V, \sigma_2 >} \quad \text{[neBig]}$$

$$\frac{< E_1, \sigma > \Rightarrow < n_1, \sigma_1 > \; < E_2, \sigma_1 > \Rightarrow < n_2, \sigma_2 > \; \mathtt{subOp}(n_1, n_2) \triangleright V}{< \mathsf{sub}(E_1, E_2), \sigma > \Rightarrow < V, \sigma_2 >} \quad \text{[subBig]}$$

$$\frac{\mathtt{valueOp}(\sigma, R) \triangleright V}{< \mathsf{deref}(R), \sigma > \Rightarrow < V, \sigma >} \quad \text{[variable]}$$

The result of running the eSOS interpreter with these rules on the input term above is a 30-step reduction of the initial term to the terminating value done, the last four configurations of which are:

```
< seq(done, assign(gcd, deref(a))), sig = { a->3 b->3 } >
< assign(gcd, deref(a)), sig = { a->3 b->3 } >
< assign(gcd, 3), sig = { a->3 b->3 } >
< done, sig = { a->3 b->3 gcd->3 } >
```

Happily, after the final step the store in the final configuration contains a binding from gcd to 3.

One can write specifications that are incomplete, but appear to work. The characteristic symptom is that the behaviour of the interpreter is sensitive to the order of the rules. In fact this specification contains nondeterminsim: rules [assign] and [assignResolve] can both trigger if the redex is an integer.

With the ordering shown here, the interpreter prioritises [assign] over [assign-Resolve] which has the effect of invoking [assignResolve] on complex expressions until they are reduced to an integer, at which point [assign] performs the assignment. If the order of the rules is reversed, the interpreter will loop forever on [assignResolve].

The cure for this class of problem is to ensure that sufficient side-conditions are added to the rules to ensure that at most one rule at a time can be triggered.

---

4   The eSOS Interpreter

The origins of the eSOS tool lie in providing efficient interpretation of rules for funcons. The software was developed within the PLanCompS project as a sort-of 'assembly language' for SOS rules. The intention was to reduce SOS rule interpretation to a minimalist core, with richer and more expressive forms of specification languages (such as Mosses' CBS notation) being translated down into eSOS before execution. Once developed, we created experimental lab sessions for the SLE course. These were very successful and led to a reworking of the course which put SOS at its centre.

In the literature a variety of notations are used within SOS specifications. Some are just syntactic sugar: for instance a turnstile symbol $\vdash$ may be used in expressions such as $\rho \vdash \langle \theta, \sigma \rangle \rightarrow \langle \theta', \sigma' \rangle$ as shorthand for $\langle \theta, \rho, \sigma \rangle \rightarrow \langle \theta', \rho, \sigma' \rangle$.

More significantly, most authors use standard mathematical notation where possible, and allow computations and function calls to appear directly within transitions. For instance, a rule for subtraction might be written:

$$\frac{n_1 \in \mathbf{Z} \quad n_2 \in \mathbf{Z}}{\langle \mathsf{sub}(n_1, n_2) \rangle \rightarrow \langle n_1 - n_2 \rangle} \qquad \text{[subConcise]}$$

using standard symbols for set membership and the set of integers. The expression in the right-hand side of the conclusion should be read as the arithmetic result of performing subtraction on the substituted variables n_1 and n_2.

These conventions certainly allow for more succinct expression, but can be a little daunting at first encounter, especially the ellision of side conditions into transitions. We might think of them as 'high level' formats which are convenient for the human reader when exercising small example specifications.

The eSOS format is extremely limited, but no less expressive than these richer forms. We can view it as a low level format in which the operations needed for our style of interpretation are explicit. eSOS allows only the three operations: pattern matching, substitution and evaluation of functions from term(s) to term. In fact the substitution operator is automatically applied to the right hand side of all transitions and side conditions, and so never needs to be written. In addition, configurations must be comprised of terms with no embedded functions.

Functions can only appear on the left hand side of side-conditions. The arguments to, and the return value from, a function, must be terms. This means that terms such as the number 67 or the boolean false are represented as trees containing a single node which is labeled with 67 or false accordingly.

New values may be computed and inserted into the result of a transition by matching the result of function to a variable, and then binding that variable in the right hand side of a conclusion, as shown in rule [sub] above.

The current eSOS interpreter works greedily in the sense that the first rule that succeeds will be used, and rules are checked in the order that they are written. Within a rule, conditions are checked in strict left to right order. In principle we could also use more sophisticated interpretation strategies that supported non-determinism so as to model concurrency.

eSOS provides a *value system* which has built in dynamic type checking allowing a designer to test parts of their implementation before they have implemented the static semantics of their type system. The system has a fixed set of operations with suggestive names such as add, union and so on. The value classes are all extensions of class Value, which contains a method for each operation. Within Value, the operation methods all throw a Java exception. The idea is that the class for, say, Integer extends Value and implements its own overriding method for each operation that is meaningful on that type. If an operation is called on an inappropriate value (for which is no operation defined) the top level method in Value will issue a run time error.

Function calls in eSOS side conditions are almost all direct calls to the functions in the value library; all the interpreter needs to do is to extract the label from a term (which will be an instance of a value package class) and call the corresponding method. The interpreter contains a case statement which branches on the function name and performs the extract-and-call action. Here is the branch for the subOp() function used in our GCD rules:

```
case "subOp":
  functionResult = new ValueTerm(
    leftPayload.sub(children.get(1).getPayload()));
  break;
```

The value system also provides a set of coercion operations which can interconvert values where appropriate.

Most of the value classes are really wrappers on the underlying Java API class. We offer these primitive types: Boolean, Character, Integer32, IntegerArbitrary, Real64, RealArbitrary, Null and Void; and these collection types: Array, String, List, Set, Tuple, Map, Record, MapHierarchy.

The IntegerArbitrary and RealArbitrary classes support arbitrary length values. The MapHierarchy class contains a map and a reference to another MapHierarchy called the parent. If a key lookup is performed on a MapHierarchy, the search proceeds recursively through the base MapHierachy and its parents. This naturally implements nested scoping of key-value bindings. In addition there are Term and TermVariable classes that construct trees whose nodes are labeled with instances of Value types. The Term class includes pattern match and substitute operations. Some of the collection classes also have implementations of match and substitute that generalise over the terms held in the collection.

The implementation of eSOS relies heavily on the classes in the value package; for instance SOS configurations are represented by instances of the Record class and environments by instances of MapHierarchy. Terms are, of course, represented by instances of value class Term and the builtin matching and substitution methods are sensitive to instances of variables represented with the TermVariable class.

With so much of the work being done within the operation methods of the Value library, the main interpreter function may be compactly expressed. The current implementation requires some 30 lines of Java.

5   The eSOS Concrete Syntax

eSOS rules may be constructed directly by programs written in Java and other JVM languages through an Application Programmer Interface (API), but the usual way to create a specification is via a text file containing eSOS concrete rules. The prelude and a concrete form of the first two rules from our GCD specification is shown below. From this, LATEX source to typeset the equations is automatically generated.

```
relation ->, sig:map, done
relation =>, sig:map
latex sig "\\sigma", -> "\\rightarrow", => "\\Rightarrow"

-sequenceDone
---
seq(done, C) -> C

-sequence
C_1 -> C_1'
---
seq(C_1, C_2) -> seq(C_1', C_2)
```

The **relation** directive declares each transition symbol and zero or more associated syntactic entities. These are typed as one of the five classes of entity mentioned on page page 6; in this case entity **sig** is of type **map** and thus can be used to model the store. The configurations of the complete specification is the union of the entities declared in all of the **relation** directives.

The **latex** directive creates a set of mappings which are used to generate LATEX aliases, enabling us to write **sig** in the source file and have it appear as σ in the typeset output.

The rules themselves are *elided* in that entities which are used in 'standard' ways need not be mentioned. This approach is inspired by Peter Mosses' work on MSOS [5], in which semantic entities are gathered into a record which labels the transition. Mosses provides a category-theoretic classification of propagation rules for entities. In eSOS we use a single propagation rule which we call the 'round the clock' rule, so for instance an unmentioned store entity σ propagates as:

$$\frac{\langle, \sigma_0 \rangle \rightarrow \langle, \sigma_1 \rangle \ \langle, \sigma_1 \rangle \rightarrow \langle, \sigma_2 \rangle \ \ldots \langle, \sigma_{k-1} \rangle \rightarrow \langle, \sigma_k \rangle}{\langle, \sigma_0 \rangle \rightarrow \langle, \sigma_k \rangle}$$

Apart from reducing the amount of writing required, the main purpose of this elision is to support modularity, allowing fragments of specifications which may use different configurations to be brought together in the manner of MSOS. Our uniform propagation rule has the merit of simplicity but in general will generate more bindings during interpretation than strictly necessary.

Space precludes a detailed example, but the motivation for adopting this capability is to support the Funcon methodology mentioned in Sect. 1. In particular, we wish to support the use of signal entities which manage the propagation of exceptions and other forms of unusual control flow. In general, the only constructs needing to access signal entities are those originating or handling the exceptions. We do not want to clutter all of the other rules with references to signals; in eSOS they can be simply elided away in the source, and will then be automatically generated as the rules are expanded for interpretation.

6   Connecting Parsers to eSOS

A BNF context free grammar for the GCD language in Sect. 2 is shown below. Terminals are stropped `'thus'` and we assume the availability of two lexical items `INTEGER` and `ID` which match decimal digit sequences and alpha-numeric identifiers in the conventional way. (Ignore for the moment the ^ annotations.) For compactness, the grammar only provides definitions for the $>$, $\neq$ and subtraction operations though it does encode their relative priorities and associativities. The grammar does not generate empty programs.

```
statement ::= seq^^ | assign^^ | if^^ | while^^
seq ::= statement statement
assign ::= ID ':='^ subExpr ';'^
if ::= 'if'^ relExpr statement 'else'^ statement
while ::= 'while'^ relExpr statement
relExpr ::= subExpr^^ | gt^^ | ne^^
gt ::= relExpr '>'^ subExpr
ne ::= relExpr '!='^ subExpr
subExpr ::= operand^^ | sub^^
sub ::= subExpr '-'^ operand
operand ::= deref^^ | INTEGER^^ | '('^ subExpr^^ ')'^
deref ::= ID
```

When used to parse the GCD program above, this grammar yields a derivation tree containing 92 nodes. The relatively large structure contains nodes representing, for instance, keywords and punctuation that may be safely discarded without losing the underlying meaning of the program. It is conventional in formal semantics work (and indeed in compiler construction) to generate a more compact intermediate form. For instance, the GNU compilers use the GENERIC libraries to build simplified trees which are translated into three-address code for optimisation, and the metamodelling community typically use Java classes to represent semantic entities which are initialised by concrete parsers.

In formal semantics, connections to concrete parsing are often eschewed in favour of starting with some abstract syntax capturing syntactic-categories such as declarations, commands, expressions and so on. This is reasonable for research, but can be a bar to progress for those wishing to simply execute semantic specifications, whether on paper or via interpreters. For example, how should phrases in the simplified abstract syntax to be constructed from a concrete program source?

An approach we have found useful is to deploy *folds* [2] to convert full derivation trees to simplified abstract syntax trees. There are two fold operations denoted by ˆ (fold under) and ˆˆ (fold over). In both cases, the annotated node A is combined with its parent P and the children of A are 'pulled up' and inserted as children of P, in order, between the siblings of A. When folding-over, the label of P is replaced by the label of A.

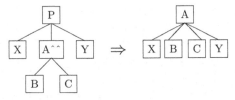

When folding under, P retains its original label and thus A disappears: a fold-under applied to a terminal, therefore, has the effect of deleting it from the tree and can be used to remove syntactic clutter such as the ' (' and ') ' terminals in the GCD grammar. Fold-overs can be used to telescope chains of nonterminals: for instance we use it above to overwrite all instances of nonterminal **operand** with **deref**, **subExpr** or an integer literal as appropriate. We have also used carrier nonterminals such as **ge** and **sub** to replace concrete syntax operators such as **>=** with alphanumeric names. The reader may like to check that the annotations above, when applied to the derivation tree for our GCD program yields this abstracted tree, which has 39 nodes.

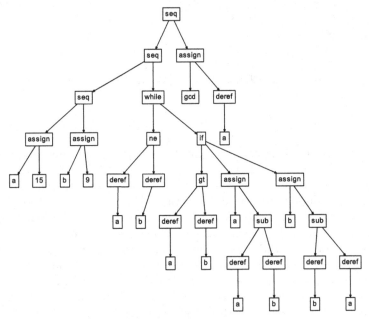

If we output the labels in a preorder traversal using the usual bracketing convention, we get this text rendition which is in a format suitable for use directly as a program term with the eSOS interpreter.

```
seq(seq(seq(assign(a, 15), assign(b, 9)),
  while(ne(deref(a), deref(b)), if(gt(deref(a), deref(b)),
    assign(a, sub(deref(a), deref(b))),
    assign(b, sub(deref(b), deref(a)))))),
  assign(gcd, deref(a)))
```

Tree construction with fold operations may be described using an L-attributed grammar and hence folded derivation trees may be produced in a single pass, or even 'on the fly' by recursive descent parsers.

7   Student Response and Conclusions

eSOS is a distillation of the core operating principles of a sequential SOS interpreter, and as such it represents a 'lowest common denominator' of the various enriched notations that one encounters in the research literature. The simple syntax, combined with a compact interpreter written in Java provide a comfortable entry to formal semantics for undergraduate students. Student response has been enthusiastic. The even split between laboratory and lecture room time enabled impressive project work, and in formal questionnaire returns students rated the course as being significantly more intellectually stimulating than the mean scores across all other courses in our school.

References

1. Churchill, M., Mosses, P.D., Sculthorpe, N., Torrini, P.: Reusable components of semantic specifications. Trans. Asp. Oriented Softw. Dev. **12**, 132–179 (2015)
2. Johnstone, A., Scott, E.: Tear-Insert-Fold grammars. In: Proceedings of the Tenth Workshop on Language Descriptions, Tools and Applications, LDTA 2010, pp. 6:1–6:8. ACM, New York (2010)
3. Kats, L.C., Visser, E.: The spoofax language workbench: rules for declarative specification of languages and ides. SIGPLAN Not. **45**(10), 444–463 (2010)
4. Klint, P., van der Storm, T., Vinju, J.J.: RASCAL: a domain specific language for source code analysis and manipulation. In: Ninth IEEE International Working Conference on Source Code Analysis and Manipulation, SCAM 2009, Edmonton, Alberta, Canada, 20–21 September 2009, pp. 168–177 (2009)
5. Mosses, P.D., New, M.J.: Implicit propagation in structural operational semantics. Electr. Notes Theor. Comput. Sci. **229**(4), 49–66 (2009)
6. Plotkin, G.D.: A structural approach to operational semantics. J. Log. Algebr. Program. **60–61**, 17–139 (2004)
7. Roşu, G., Şerbănuţă, T.F.: An overview of the K semantic framework. J. Log. Algebr. Program. **79**(6), 397–434 (2010)
8. Scott, E., Johnstone, A.: GLL syntax analysers for EBNF grammars. Sci. Comput. Program. **166**, 120–145 (2018)

9. Sewell, P., et al.: Ott: effective tool support for the working semanticist. J. Funct. Program. **20**(1), 71–122 (2010)
10. van den Brand, M., Heering, J., Klint, P., Olivier, P.: Compiling language definitions: the ASF+SDF compiler. ACM Trans. Program. Lang. Syst. **24**(4), 334–368 (2002)

Managing Heterogeneity and Bridging the Gap in Teaching Formal Methods

Pamela Fleischmann, Mitja Kulczynski[(⊠)], Dirk Nowotka, and Thomas Wilke

Department of Computer Science, Kiel University, Kiel, Germany
{fpa,mku,dn}@informatik.uni-kiel.de, thomas.wilke@email.uni-kiel.de

Abstract. At Kiel University, a course on theory of computation and a course on logic in computer science form the basis for teaching formal methods. Over the last years, new study programmes (computer science teacher training, business information systems, computer science for international students on master level) have been established, calling for changes to the courses. Guided by the experience gathered over time, course syllabi as well as teaching and examination formats and practices were adapted, resulting in a complex scheme. In this paper, we review this development, with particular focus on managing heterogeneity and bridging the gap between actual and required qualification of enrolling students. Key ingredients of our teaching methods are a spiral approach, frequent testing, supervised learning time, and a game.

1 Introduction

The department of CS at Kiel University enrolls most of the students for its bachelor's programme in CS; many of the students finishing this programme are then enrolled to the consecutive master's programme. In both programmes, several electives on using formal methods, relating to formal methods and on formal methods directly are offered: Introduction to Formal Software Analysis, Algebraic Specification, Engineering Secure Software Systems, Decision Problems, and so on. The basis for all these courses is laid in the bachelor's programme, where a fundamental course on theory of computation (ToC) and one on logic in computer science (LiCS) are compulsory. These courses build, in turn, on three math courses. The five courses altogether are worth 40 credits.

Resources at Kiel University are scarce, in particular, the staffing level is low, both in terms of professors and scientific staff. So, as new study programmes have emerged over time, the ToC and the LiCS course had to be adapted to the different groups of students: in addition to the students from the aforementioned bachelor's programme in CS, students from a business information systems (BIS) programme, a two-area (2A) programme with one area being CS, and a CS teacher training (CSTT) programme have to be catered for.—The department had to *manage heterogeneity*; it was no option to offer separate courses for each study group.

When, in 2017, the CS department decided to set up an international master's programme, a new problem arose with regard to ToC and LiCS. The students

© Springer Nature Switzerland AG 2019
B. Dongol et al. (Eds.): FMTea 2019, LNCS 11758, pp. 181–195, 2019.
https://doi.org/10.1007/978-3-030-32441-4_12

who apply to this programme are not fully qualified according to our regulations: they are often missing modules in theory, logic, and math. In addition, their learning habits are very different from the ones our students have acquired. One reason might be completely different examination practices: we often give our students a few complex problems which they have to work on for some time; the students applying to us are used to exams with many questions which need to be answered in brief and quickly. So the CS department had to somehow *bridge the gap* between the actual qualification of the admitted students and the required qualification and help the students get accustomed to the local teaching, learning, and examination culture.

In this paper, we describe how the department adapted ToC and LiCS to the changing requirements, which experiences were gathered, and how these experiences induced further changes. To be more precise, we describe the *interplay* between changing requirements, experiences gathered, and modifications made. The key insights that we have gained are:

1. A *spiral approach* (see [4]) has been successful for coping with heterogeneity.
2. *Frequent testing* (see [6,10]) as an examination practice has been effective in bridging the gap and making the students accustomed to the local teaching and learning culture.
3. In general, *supervised learning time* is effective and economical.

We hope our experiences may help other CS departments when confronted with similar problems.

Structure of the Paper. The next section provides details about the study programmes offered by the CS department of Kiel University and their history. There are two main sections, one about managing heterogeneity (Sect. 3) and one about bridging the gap (Sect. 4), and there is a final section (Sect. 5) which provides statistical data and an evaluation of our teaching approaches.

2   The Study Programmes and Their History in Short

In the following, we describe the study programmes offered by the CS department of Kiel University briefly. The formal examination regulations can be found at [5].

2.1   Computer Science

When, in 2002, the CS department of Kiel University replaced the diploma study programme in CS by a bachelor's and a consecutive master's programme, it established two courses as the basis for teaching formal methods: a course on ToC and another one on LiCS, both placed in the second year of the bachelor's programme, each of them spanning 14 weeks with 180 min (four hours) teaching and 90 min (two hours) classroom exercises per week (worth eight credits in the ECTS [9]). Roughly, the theory course covered Chaps. 1 through 5 of [13] and fundamental notions and facts of complexity theory, while the logic course covered Chaps. 1 through 12, except 3, 5, and 8, of [2]. Four math courses, which addressed math students in the first place, were part of the first two years of the bachelor's programme.

In 2008, the four math courses were replaced by three newly designed math courses (Mth-A, Mth-B, Mth-C), directly addressing students of computer science. ToC and LiCS were moved to the end of the second year and the beginning of the third year, respectively, following the three math courses. In addition, an advanced programming course (AP), which offered an introduction into logic programming, was introduced.

2.2 Business Information Systems

In 2006, the CS department of Kiel University established a bachelor's and a consecutive master's programme in business information systems. Formal methods are relevant to this program when it comes to business process models. In contrast to the CS students, students from BIS take only two math courses (Mth-A and Mth-B).

ToC has been a substantial ingredient of the BIS study programmes, originally located in the master's programme, then reduced to half the credits (ToCBIS), and later on moved to Semester 4 of the bachelor's programme.

There were no resources for installing a separate course. Section 3 describes the heterogeneous teaching concept we developed.

2.3 2-Area Programme

In 2014, Kiel University established two university-wide so-called two-area bachelor's and master's degrees, one of them being a high-school teacher degree, the other one aiming at the general job market. The CS department made Mth-A and ToC compulsory courses for the bachelor's CS area. Later on, the number of the CS-specific courses was increased at the expense of Mth-A, resulting in the need to offer an adapted theory course (ToC2A).

Just as with BIS, there were no resources for installing a separate course. Section 3 describes the heterogeneous teaching concept we developed.

2.4 International Programme

For the first time in winter 2017/18, the CS department offered a master's programme in English, addressing, in the first place, students from abroad. As a reaction to the fact that many applicants were missing a thorough education in theory and logic, the department took several different measures to address this, see Sect. 4.

3 Managing Heterogeneity

The ToC course and the LiCS course were originally designed for the bachelor's programme in CS and the consecutive master's programme only. They were based on a considerate amount of math and supposed to prepare for advanced courses on formal methods. They were meant to convey core concepts in theory and logic.

These days, the ToC course serves a heterogeneous body of students with diverse study objectives and prior knowledge:

1. students from the bachelor's programme in CS
   > ToC is compulsory in Semester 4
   > Mth-A, Mth-B, and Mth-C are prior knowledge
2. students from the bachelor's programme in BIS
   > a smaller ToC (TocBIS) is compulsory in Semester 4
   > Mth-A, Mth-B, and Mth-C are prior knowledge
3. students from the 2A bachelor's programme
   > a smaller ToC (ToC2A) is compulsory in Semester 2
   > no math courses

3.1 Teaching Approaches for Managing Heterogeneity

In the following we review how the teaching approach with regard to ToC and LiCS has evolved over the past two years, as a reaction to the changes in the study programmes. We describe the model prior to theses changes and discuss the need for modifications as well as our experiences.

All our courses are taught in German. The ToC course covers the Chomsky hierarchy and automata, basics on theory of computation, and complexity theory. The LiCS course covers propositional and first-order logic: syntax and semantics, decidability of satisfiability, normal forms, and modelling problems in logic.

Teaching Approach Prior to 2016. The *traditional format* of these two courses is the same as that for many other courses at our department: there are fourteen weeks of teaching, where in every week there are two teacher-centered lectures of 90 min each and assisting tutorials of 90 min with about 30 students participating. All tutorials are the same; each student attends one of them. The students are given weekly homework exercises. The solutions they turn in are corrected, and good solutions are presented and discussed in the tutorials. The tutorials are conducted by undergraduate students without prior training on how to teach (but with a proper supervision by a doctoral student). As a result, the quality of the tutorials varies extremely. Each course has a final written exam at the end of the semester, consisting of problems similar to the homework exercises. These two courses had the reputation of being the toughest of the curriculum beside the introductory math course Mth-A.

ToC underwent major changes in the last two years. This is what we describe in the following.

Changes in Winter 2016/17. In winter 2016/17, ToC had to be offered in the full version (8 credits) and a reduced form (ToCBIS, 4 credits). In summer 2017, the reduced form would also be part of the 2A bachelor's programme.

Teaching Approach. We realized a spiral approach: we split the semester in half, and in the first half we covered the same topics as previously in the entire semester, but on a lower level of detail and without proofs; in the second half we covered all topics again, but on a more detailed level and with proofs, see below for details and compare Fig. 1.

The students with less prior knowledge (BIS) were only required to attend the first half of the course (and obtained only half the credits), while the students from the bachelor's programme in CS had to attend the full course.

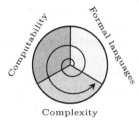

Complexity

Fig. 1. Structure of the course

In the first half, every topic was motivated with an introductory example. Then formal definitions followed, which were then enriched by more examples. Selected basic theorems were only stated and not proven, but intuition was conveyed. Simpler constructions and algorithms were presented and practiced.

In the second half, formal proofs were given for some of the theorems, more specialized theorems were dealt with, more involved concepts were introduced, and more complex constructions were presented and practiced.

Here are some details of how the part on finite-state automata was split. The first half started out with an illustration of several DFAs, followed by a formal definition of DFA, followed by more examples cast in terms of that definition. The Pumping Lemma was discussed in the first half and illustrated using the sketch given in Fig. 2. The students also learned how to apply the Pumping Lemma to prove that a language is not regular. The power set construction, being a simple construction, was also presented in the first half.

The minimization of DFAs was part of the second half. The same applies to the proof of the Pumping Lemma or the proof of correctness of the power set construction.

We redesigned the homework exercises completely. In the first half of the course we asked the students to construct automata for given languages, apply algorithms such as the power set construction or check the correctness

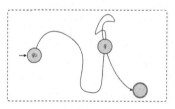

Fig. 2. Pumping lemma

of many-one reductions. Only in the second half, proofs were required.

We kept the general format of the tutorials. Unlike before, we provided the tutors with a sample solution to each problem, which they would rely on in their tutorials and which they would pass on to the students. In a tutorial session, the problems of that week were worked on and solved in the group, based on the solutions produced by the individual participants prior to the tutorial and guided by the tutor. The idea was to produce a joint solution in a collaborative way. In addition, one new problem was discussed, introducing the topic of the homework exercise for the following week.

We created two sets of problems for the end-term exam. Each of the sets consisted of a multiple choice section (reproduction of knowledge) and five further problems, each of them similar to, but in general somewhat easier than the problems in the homework exercises. Each set was set up for 1.5 h. The students that had only attended the first half of the course had to work on the first set; the other students had to work on both sets. In addition, we offered an optional mid-term exam. By passing this exam the students were released from working on the first set of problems in the final exam.

Feedback. The students' feedback to the spiral approach was mostly positive. They liked the idea of revisiting each topic and believed that this had improved their understanding. There was a debate among the students whether in the first half of the lecture the general approach had been too formal or not. They complained about the way we had covered (un-)decidability reductions, which then led us to cover this in more detail in the following semester. The increased number of examples in the lecture was considered a positive feature.

The feedback on the tutorials was mostly positive. This had not been the case in previous years. The students believed they had gained a better understanding of formal proofs. They credited us for being patient and for the interactive development of solutions in the tutorials. In some cases, even in the first half, students complained about missing mathematical skills on their own part. The mid-term test received, in general, a positive review. However, the students noted they had almost no time for preparing themselves for this test.

The overall evaluation of the tutorials and the lecture was much better than in previous years.

Changes in Summer 2017. We kept the spiral approach.—Recall that in this semester the first half (reduced form) was not only compulsory for BIS students, but also for 2A students.

Prior to this semester the LiCS course had been taught in the fourth and the ToC course in the fifth semester of the CS bachelor's programme. In 2017 the two courses switched their places, calling for changes, because the part of ToC on complexity theory had been heavily centered around satisfiability in propositional logic (SAT), a problem discussed in depth in LiCS.

So in this semester in ToC we gave (only) a brief introduction into satsfiability and reduced the part on complexity theory. We skipped several of the logic-related algorithmic problems, for instance, the quantified boolean formula problem (QBF).

Proving NP-completeness of SAT—an essential part of the second half of the course—now was treated on a rather informal level, by presenting only the tableaux idea as depicted in Fig. 3 (see [13]), but not going into details of the proof. A positive effect of this was that we had some more time to go into details of other topics: we introduced unbounded grammars as a formalism equivalent to Turing machines.

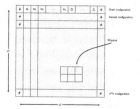

Fig. 3. Tableaux used in SAT proof

The tutorials underwent a major reorganization: we offered two four-hour slots of *supervised learning time (SLT)*. The students worked in small groups of two to four people. They mainly discussed the problems from their homework exercises, but also general problems occurring within the course. A highly qualified doctoral student guided the small groups individually. The main advantage of this tutorial style is the individual assistance by a qualified tutor. In a discussion with a small group, the tutor can easily adjust to the level of expertise of individual members of the group. A drawback of this style is that it lacks open discussions (which are suitable for general questions) and concrete instruction.

Changes in Summer 2018. As a reaction to the drawback just mentioned, we reintroduced a third tutorial following the traditional style: guided by a qualified doctoral student, the participants developed solutions to difficult problems in class. This tutorial had, in fact, a positive impact: the quality of the solutions to homework exercises turned in by students increased. In general, the feedback to the overall tutorial scheme in this semester was very positive: we did not receive any negative feedback.

Changes in Summer 2019. In the current semester, the lecture is the same. We still have not found a solution to the problem that propositional and predicate logic are treated on an informal level. The homework exercises are now augmented by programming exercises where students are asked to implement techniques and structures introduced in the lecture (e.g., a deterministic finite-state automaton, a decision procedure whether a context-free language accepts a given word, or a Turing machine). Working practically with simple structures such as DFAs is, from our point of view, the best way to get into the more complex details of formal software verification. In addition, *implementing an abstract concept* supports the understanding. The implementation is done in Python 3, a script-based language, which is in general quite self-explanatory (see [1]). The feedback, so far, has been positive.

4 Bridging the Gap

For the first time in winter 2017/18 the CS department of Kiel University offered a master's programme in English, open to students from all over the world, following the same pattern (and, in fact, according to the same examination regulations) as the master's programme offered in German.

As formal methods are part of the master's programme, a solid expertise in theory and logic is required and explains why in our bachelor's programme courses in theory of computation and logic in computer science are mandatory.

All of the students that applied to us were missing an adequate LiCS course in their bachelor's programmes and many of them were missing an adequate ToC course, which meant we had to "bridge the gap" between actual qualification and required qualification.

In what follows we describe in more detail the student body of the programme and our way to bridge the gap.

4.1 Students

Criteria for Admission. The general rule of admission is that prospective master students need to have obtained

(∗) a qualification which they would have obtained through our bachelor's programme.

In this sense, our master's programme is consecutive.

Prospective students who are not fully qualified can still be admitted, provided that their gap corresponds to a workload of credits in ECTS of at most 30 and on the condition they catch up on the missing qualification within half a year or a full year of enrollment. For details, see below.

From the general rule (∗), we derive the following concrete admission criteria. The students need to have

> a bachelor's degree in computer science or in a similar domain like computational engineering or computer and information systems,
> taken courses in mathematics worth at least 24 credits, and
> profound language skills in English (IELTS 6.0, TOEFL-ITP 550, or an English proficiency letter by a university).

Countries of Origin. More than three quarters of our international students originate from Pakistan and India; the remaining fourth is made up of students from up to 18 different countries, see Fig. 1. The latter group of students is heterogeneous with regard to their academic qualification and culture. In the following, we refer to the students beginning in winter 2017, summer 2018, ... by Cohort I, II, ...

Additional Requirements—Opening the Gap. As stated above, we have the option to admit students who are not fully qualified, under certain conditions. As no international student that has applied to us was fully qualified (in the sense of our regulations), we admitted partially qualified students. As these students are required to obtain a full qualification within their first half year or their first year, we speak of *students with additional requirements (SWARs)*.

Table 1. Countries of origins

Cohort	Pakistan	India	Bangladesh	Iran	Nigeria	Yemen	Other
I	1	1					2
II	12	6					6
III	37	21	8	2	1		6
IV	45	22	12	5	2	2	9

As we wanted (and want) to support SWARs, we admitted only students that were missing qualifications in LiCS, ToC, advanced programming (distributed programming, functional programming, logic programming), and computer and network security, because we did not have lecturers (nor examiners) for the other subjects.

Since the number of applicants which fell under this rule increased over time, see Table 2, we now only admit applicants with a missing qualification in LiCS and AP only. The reason for the particular choice of LiCS and AP is that these subjects are key ingredients of the teaching profile of our department.

Prior Knowledge—Widening the Gap. While the three courses in mathematics in our bachelor's curriculum provide all students with a thorough and rigorous basis in mathematics, in particular, all students are skilled in finding and carrying out mathematical proofs to a certain extent, our international students are trained in applied or engineering mathematics. Some of them may be able to phrase a mathematical statement like the Pumping Lemma for regular languages and apply it, but none of them can give a proof of it; some of them know how to construct a finite-state automaton for a given regular language, but none of them know what it means to prove that the automaton constructed is correct, let alone to come up with a proof themselves.

Even though we only admit students with profound language skills in English, not all international students speak and understand English sufficiently well for following lectures, participating actively in class room exercises, or taking an oral exam. In general, our international students have difficulties with abstract thinking.

Table 2. Number of applications and admissions

Semester	Applicants	Admissions	Obligations in		
			ToC	LiCS	AP
Winter 2017/18	65	52	3	6	6
Summer 2018	113	85	9	21	26
Winter 2018/19	221	183	21	59	73
Summer 2019	347	136	0	132	134

Table 3. Teaching approaches over time

	Cohort			
	I	II	III	IV
First semester	**LICS** · self-studies · oral exam	**ToC** · lecture · homework · oral exam	**LiCS/ToC** · lecture · homework · 4 written exams · HtP	**LiCS/ToC** · lecture · homework · 4 written exams · HtP · tutorial
Second Semester	**ToC** · lecture · homework · oral exam	**LiCS (ToC)** · lecture · homework · 4 written exams · HtP		

Learning Environment. First of all, the majority of our international students need to rebuild an attitude towards learning in an academic environment, as they have been employed in the job market for a couple of years when they start their master's studies with us. (This is in sharp contrast to our German students.) In addition, due to a problematic housing situation (in all over Germany), it takes most of our international students a couple of months to find a place to stay for longer. This makes their first steps even more difficult. It is, however, worth noticing that at Kiel university there are well-organized Pakistani and Indian communities welcoming and supporting new students upon arrival, whereas this is, more or less, not the case for other countries.

4.2 Teaching Schemes for Bridging the Gap

It is important to understand that our regular teaching is quite regulated. However, how we support SWARs to obtain a full qualification and verify they have been successful, is essentially unregulated: the department can install own rules.

We started with a teaching scheme we thought was reasonable and changed the scheme over time according to the experiences we gathered, but also adjusted it to the resources available to us. In general, the level of support increased steadily.

In the following, we outline our schemes, see also Table 3.

Winter 2017/18. Cohort I, which started in winter 2017/18, was aligned to the study programme of the German students: LiCS in winter and ToC in summer. This meant Cohort I had one year time to fulfill their obligations: LiCS in the first semester; ToC in the second semester.

For LiCS, we offered a weekly tutorial of two hours. The students were expected to prepare a section from [12] through self-studying for the tutorial. During the tutorial, first the section was discussed. Then the students were given the opportunity to work on a set of problems and discuss these problems.

The active participation in this tutorial was poor; only a few students showed up. It was almost impossible for the students to solve the problems given to them. Interestingly, it was almost impossible for them to describe the difficulties they encountered when trying to solve the problems.

In February, at the end of the semester, the students had to pass oral exams of 30 min. The questions (if understood at all) were answered only very vaguely and not with the precision one would expect for a mathematically rigorous course.

We concluded that self-studies with weekly discussions are not sufficient for learning LiCS for students lacking abstract thinking and appropriate mathematical skills. This led to minor changes for the summer 2018, when the same cohort, together with the new one, Cohort II, was supposed to take ToC.

Summer 2018. Cohort II, just as Cohort I, was given one year to clear their obligations: in summer 2018, they were supposed to take ToC; in the following winter, they were supposed to take LiCS (and, potentially, repeat ToC if they had failed it in their first attempt).

In contrast to the previous semester we offered a lecture of two hours once a week based on [8], i.e. no self-studies anymore, but a lecture and an accompanying textbook. In addition, the students were given homework exercises with a processing time of one week. Their solutions were corrected very carefully: the students were given hints to what was wrong and advice on how they could improve. Moreover, the students were provided with sample solutions.

While the number of students who worked on the assignments dropped as the semester went on (we have the same experience with the German students), the students participated actively and continuously in the lecture.

The solutions submitted by the students reflected our observations from the previous semester: as long as the problems were very similar to the examples from the lecture and did not involve proofs, the results were fine; but the students were neither able to phrase a decent proof nor to transfer what they had learnt to new situations. Even though the absolute results of the students were better during the semester, only half of the students passed the oral exam at the end of the semester in July. To be very clear, the reason was that they had no experience on how to work mathematically rigorously.

Winter 2018/19. Since winter 2018/19 we have offered, in each semester, a LiCS and (!) a ToC course. This means, in particular, that from Cohort III onward each cohort has only one semester to clear their obligations (which explains the two empty entries in Table 3). There were two reasons for this change: first, this way we can handle the increased number of students better; second, the students get an earlier feedback whether they are allowed to pursue their master's studies.

In addition, we changed our format of examination drastically. None of the students of Cohorts I and II had had an oral exam in their home countries. So we switched to written exams. Following the frequent test paradigm, see [10], the students had to pass three out of four one-hour long exams, with two attempts at each exam. The exams were spread over the entire semester, see Table 4.

Table 4. Examination schedule during a semester where Ex a.b represents the bth attempt of the ath exam.

Date 1	Date 2	Date 3	Date 4	Date 5
Ex 1.1	Ex 2.1	Ex 3.1	Ex 4.1	
	Ex 1.2	Ex 2.2	Ex 3.2	Ex 4.2

In practice, we deviated from the scheme on one occasion: the exam on date 5 was oral, because only four students took the exams.

The teaching remained the same as in the previous semester: two hours lecture per week and corrected homework exercises supplemented by sample solutions. The homework was designed in the same way as the exams. Each assignment consisted of four parts:

1. *true/false* part: determine whether given statements are true or false.
2. *definition* part: rephrase definitions of formal notions
3. *thinking* part: solve small problems, in particular, prove statements on your own
4. *reversed learning* part (only in homework): create one (!) true/false-question on your own

The definition part was crucial, after we had noticed that too many students had only a rough and superficial idea of the formal notions in the lecture and thus could not work properly with them. In particular, they had not been able to solve problems from the first and the third part. We experienced that now the students performed better at the first and the third part.

The fourth part gave us a relaxed way to start the next lecture with a *game*: each student passes on his/her question to some other student; this students tries to answers the question on the same sheet of paper and passes it on to a third student; this third student reads the question aloud and has to decide whether the answer is correct. See [3].

The students seemed to spend a lot of time preparing a question for the game that would be answered incorrectly. Although the game was meant to be anonymous the authors of the questions often revealed their identity as they were eager to communicate what they had intended with their question. It is worth noting that the students who created the question had a good understanding of the respective issue. An incentive for participating in the game is that three out of the ten questions in the exams are former questions by the students.

We also changed the textbook to [7]. Our previous choice, [12], was based on the good experience we had with the German version of [11]. The English translation turned out not to work as well.

The final change in our teaching scheme was a new small lecture: we offered a lecture *How to Prove? Mathematical Techniques for Proving* (HtP) based on our observations with the first two cohorts. The students had problems with abstract thinking and basic notions of mathematics such as sets and functions, as well as basic proof techniques. Consequently, the lecture covered exactly this, and, in addition, it went into details of problems discovered in the LiCS and the ToC course.

Summer 2019. Due to more resources we are able to offer an additional tutorial of 90 min per week. Based on the good experience with SLT (see Sect. 3) we decided to offer this tutorial as SLT. Until the writing of this paper, this tutorial has had a high number of participants and received good feedback.

5   Quantitative Data and Evaluation

5.1   Managing Heterogeneity

In Table 5, the number of participants and the pass rates for three consecutive ToC courses are displayed, along with the changes to the respective teaching approach. By comparison, the pass rates for prior ToC courses—the students taking these courses formed a homogeneous group—were between 40 and 50 %. The bottom line is that our current approach to managing heterogeneity yields better results than the traditional one obtained with a homogeneous group.

Obviously, the drop-out rates are unsatisfactory. This is something we have not been able to investigate so far.

Table 5. Pass rates for ToC

Semester	Change	Students registered	Exams taken	Success	Pass rate %
Winter 16/17	Spiral approach	115	100	28	28
Summer 17	Supervised learning group etc.	180	47	25	53
Summer 18	Additional tutorial	218	77	51	66

5.2   Bridging the Gap

In Cohort I, which started in winter 2017, three students had ToC and 6 students had LiCS as an additional requirement. One student passed the ToC exam and three passed the LiCS exam (33% ToC, 50% LiCS). Note that LiCS was offered in winter 2017, whereas ToC was offered in summer 2018, which means the formats were different.

In Cohort II, which started in summer 2018, nine students had ToC and 21 students had LiCS as an additional requirement. The rate of passing increased significantly after we had replaced self-studies by a lecture: five students passed the ToC exams and 14 students passed the LiCS exams (56% ToC, 67% LiCS). Note that ToC was offered in summer 2018, whereas LiCS was offered in winter 2018, which, again, means the formats were different.

Frequent testing, introduced in winter 2018, had only a major impact on ToC: 15 out of 21 students passed ToC and 41 out of 59 students passed LiCS (71% ToC, 69% LiCS). Notice that these students had twice the workload compared to the previous cohorts.

See also Table 6; recall the different teaching schedules summed up in Table 3.

Table 6. Success rates for pre-master courses

Cohort	Pass rate ToC	Pass rate LiCS
I	33%	50%
II	56%	67%
III	69%	71%

The first observation is that there is a steady increase in the pass rates.

The second observation is that frequent testing had a much higher impact for ToC than for LiCS. This may be interpreted as follows. ToC requires much more a continuous work throughout the semester.

The third observation is that, in general, the results for ToC are worse than for LiCS, even though the groups were almost identical, the lecturer was the same, etc. We offer two potential explanations: ToC relies on more mathematical tools (feedback of the students); ToC intertwines mathematics and computer science much more than LiCS.

All students who participated in the voluntary lecture HtP passed the exams in the first try.

6   Conclusion

Laying the foundations for teaching formal methods is a complex and tedious task. The experiences we have gathered show:

1. *Frequent testing* increases the pass rates, even when a tight schedule with overlapping exams (second attempt on the same date as the next exam) is used.
2. *Supervised learning time* is especially suited to teach a mathematically rigorous subject (early feedback and small-step guidance necessary).
3. A *spiral approach* (here used in the context of heterogeneity) can still be applied when used within a short period of time (one semester), and helps to save resources. In addition, it is positively received by the students!

The experience we have gathered so far makes us change (improve!) our teaching scheme for the international students again for the cohort starting this coming winter: we will introduce a spiral approach in the pre-master course. More precisely, we will offer one combined course—*Logic and Theoretical Foundations of Computer Science*—with a spiral curriculum. We will start with propositional logic (logic), go on to a part on simple mathematical proofs and proof techniques (math) and on natural proofs (logic), to automata and the Chomsky hierarchy (theory of computation), to predicate logic (logic), to more complex mathematical proofs (math), and, finally, to complexity theory (theory of computation).— This should lay the right foundation for learning formal methods in the master's programme.

Acknowledgment. We thank Andreas Mühling for comments on drafts of this paper and Michael Hanus for providing us with more details of the history of our study programmes.

References

1. Python 3.0. https://www.python.org/download/releases/3.0/
2. Ben-Ari, M.: Mathematical logic for computer science. Prentice-Hall International Series in Computer Science. Prentice-Hall International, Hemel Hempstead (1993)
3. Black, P., Wiliam, D.: Assessment and classroom learning. Assess. Educ. Principles Policy Pract. **5**(1), 7–74 (1998)
4. J.S. Bruner and National Academy of Sciences (U.S.). The Process of Education. Harvard University Press, Cambridge (1977)
5. Kiel University Department of Computer Science. Examination regulations in computer science (1999). https://www.inf.uni-kiel.de/en/studies/examinations/examination-regulations
6. Hattie, J.: Visible Learning. Routledge, London (2008)
7. Huth, M., Ryan, M.: Logic in Computer Science: Modelling and Reasoning About Systems. Cambridge University Press, Cambridge (2000)
8. Kozen, D.: Automata and Computability. Springer, New York (1997). https://doi.org/10.1007/978-1-4612-1844-9
9. Publications Office of the European Union. ECTS users' guide (2015)
10. Schneider, M., Preckel, F.: Variables associated with achievement in higher education. A systematic review of meta-analyses. Psychol. Bull. **143**(6), 565–600 (2017)
11. Schöning, U.: Logik für Informatiker. BI Wissenschaftsverlag, Mannheim (1987)
12. Schöning, U.: Logic for Computer Scientists. Birkhäuser, Boston (1989)
13. Sipser, M.: Introduction to the Theory of Computation. PWS, Boston (1996)

Teaching Introductory Formal Methods and Discrete Mathematics to Software Engineers: Reflections on a Modelling-Focussed Approach

Andrew Simpson[(✉)]

Department of Computer Science, University of Oxford,
Wolfson Building, Parks Road, Oxford OX1 3QD, UK
andrew.simpson@cs.ox.ac.uk

Abstract. Much has been written about the challenges of teaching discrete mathematics and formal methods. In this paper we discuss the experiences of delivering a course that serves as an introduction to both. The one-week intensive course, *Software Engineering Mathematics*, is delivered as part of the University of Oxford's Software Engineering Programme to groups of professional software and security engineers studying for master's degrees on a part-time basis. We describe how a change in the course's emphasis—involving a shift towards a focus on modelling-based group exercises—has given rise to some pleasing results.

1 Introduction

Much has been written about the difficulties of teaching discrete mathematics and formal methods, with problems associated with 'getting' abstraction, student motivation and what might be termed 'math-phobia' being recurring themes. Proposed solutions include the utilisation of a 'stealth-like' approach [19] ("we sneak up on our blissfully unaware students, slip a dose of formal methods into their coursework and development environments, then with a thunderclap disappear in a puff of smoke" [19]), a clear justification [29], a considered approach to links with the rest of the curriculum [30], and a focus on modelling [4].

Our focus in this paper is a one-week intensive course, *Software Engineering Mathematics*, which is delivered as part of the University of Oxford's Software Engineering Programme[1] to groups of professional software and security engineers who are studying for master's degrees on a part-time basis. The course aims to do two things: introduce students to formal methods and teach them core discrete mathematics concepts (in a fashion similar to, for example, the courses described by Warford [39] and Jaume and Laurent [18]).

Teaching part-time students who are predominantly drawn from the software engineering industry has its advantages when compared to teaching full-time

---

[1] http://www.cs.ox.ac.uk/softeng/.

© Springer Nature Switzerland AG 2019
B. Dongol et al. (Eds.): FMTea 2019, LNCS 11758, pp. 196–211, 2019.
https://doi.org/10.1007/978-3-030-32441-4_13

undergraduate students—such students bring 'real-world' experience and problems to the classroom, which helps those delivering courses to make connections between theory and practice, and to demonstrate potential benefits. In addition, such students tend to be very motivated—the financial and time investments required are, after all, significant. (The overall course costs are approximately £25K. In addition, the students are required to spend 11 weeks in Oxford, and commit several hundreds of hours to assignments and project work.) On the other hand, there are complexities associated with teaching such students: the diversity of prior academic and industrial experience, as well as a diversity of expectations, can make for an extremely heterogeneous mix of participants. A further challenge involves demonstrating that an appropriate application of the techniques being taught is relevant to the students' everyday activity—and, as such, justifies the aforementioned investments.

Of course, the difficulty of demonstrating the 'pay-off' of many Computer Science and Software Engineering tools and techniques is a challenge that has been recognised widely. For example, to quote Finkelstein [10]:

"Software engineering is, in large part, about scale. Illuminating the essence of a software engineering technique and motivating the students with convincing arguments for its value, without giving examples which are so large as to submerge the student in extraneous detail is extremely difficult." [10]

The philosophy of the course under consideration in this paper is sympathetic to the view that an 'appropriate' and 'within context' application of formal and mathematical techniques is essential to demonstrating their potential value to professional software engineers. In many ways, this is consistent with the argument put forward by Woodcock et al. [42]:

"One of the main difficulties in engineering is the cost-effective choice of what to do and where. No engineer gives the same attention to all the rivets: those below the waterline are singled out; similarly, a formalism need not be applied in full depth to all components of an entire product and through all stages of their development, and that is what we see in practice." [42]

In this paper, we show how a change to the emphasis of our Software Engineering Mathematics course—involving a shift towards a heavy focus on modelling-based group exercises—has given rise to some positive results. Our aims have much in common with those of Larsen et al. who, in [23], describe "experiences developing and delivering courses that endeavour to equip students with generic skills of abstraction and rigorous analysis by means of lightweight formal methods using VDM and its support tools." Further, our journey has much in common with that described by Cowling [3]:

"The starting point for this experience was the approach of teaching Z as a formal specification method, as presented in the standard textbooks. The problem that was soon found with this approach was that these texts did not suggest any method for constructing specifications, but instead focused on the various mathematical constructions that could be employed in the specifications. This focus left the students feeling a bit like the audience at a magic show, asking the question 'where did that bit of the specification come from', meaning that they were gaining little understanding of how they could actually use such methods themselves." [3]

In Sect. 2 we briefly consider related work. Then, in Sect. 3, we discuss the context of the contribution. In Sect. 4, we reflect upon how experience led us to the restructured version of the course that we now use. In Sect. 5 we present some indicative (and caveated) results. Finally, we conclude in Sect. 6.

2 Related Work

Our focus is a course that exists at the academic–industry interface. This is an area covered by a number of authors, including Mead *et al.* [26], Fraser *et al.* [11], Vaughn and Carver [38], Subrahmanyam [36], and Almi *et al.* [1]. In addition, in a series of papers [13,14,28], Taguchi and colleagues discuss their experiences of educating Software Engineering professionals in Japan.

The importance of abstraction and modelling[2] to the practising software engineeri is recognised widely ("We all know that the only mental tool by means of which a very finite piece of reasoning can cover a myriad cases is called 'abstraction'; as a result the effective exploitation of his powers of abstraction must be regarded as one of the most vital activities of a competent programmer" [8]; see also [7] and [40]); the difficulties of teaching abstraction and modelling is also acknowledged by many [16,20,22]. To quote Fincher and Utting [9]:

> "we know that abstraction is a very difficult step to take ... that learners find it difficult to grasp the principles embodied in a single example (or a series of single examples) then isolate it as the common referent they all share (that is, abstract from the details to the principle) and apply that principle in novel situations." [9]

Addressing these challenges is at the heart of this paper.

The course under consideration in this paper leverages the mathematical language of Z [35,41], and our contribution discusses the value of case studies. It is worth recognising that there is a rich history of Z case studies: running from the early contributions of the likes of Hayes, Morgan and Sufrin [12,27], through Jacky's *The Way of Z* [17], to more recent contributions such as [37].

3 Context

We now consider the context of the course. We start by discussing the Software Engineering Programme at the University of Oxford and then turn our attention to the Software Engineering Mathematics course.

---

[2] We would argue that, in this context, at least, the two go hand-in-hand—a 'complementary partnership' in the words of Kramer [21].

3.1   The Software Engineering Programme

The Software Engineering Programme at the University of Oxford, which was established in the early 1980s, built on the University Oxford's experience in delivering one-week intensive courses to professional software engineers in, for example, formal methods such as Z [35] and CSP [31]. An 'integrated programme' of six one-week modules was established in 1993; the Software Engineering Programme now offers one-week courses in over 40 topics. The programme also offers students the opportunity to study for MScs in Software Engineering and Software and Systems Security on a part-time basis.

At present, approximately 300 students are registered with the Software Engineering Programme. Students are drawn from a wide range of backgrounds, including large IT firms, government organisations, small companies, and the financial sector. The programme's requirements for entry are flexible, taking into account prior industrial experience, as well as academic background.

The wide diversity of the student body gives rise to a number of challenges: few assumptions can be made about the nature of previous industrial and academic experience, meaning that the complexities of teaching modelling techniques are slightly different to those associated with teaching cohorts of full-time student that are (typically) more homogeneous. Some of those complexities are discussed in [33], and it is worth reprising those arguments here:

> "The typical student on the Software Engineering Programme 20 years ago was a relatively experienced software engineer, who had been based in the industry for at least five years. This meant that the prior knowledge that one might use in delivering courses was relatively uniform. As an example, when teaching discrete mathematics, one might use a binary tree as a motivating example when discussing recursive functions. Unfortunately, this is no longer true: it is not unusual to be met by blank faces (by even those with a first degree in an IT-related subject) when mentioning binary trees. This is for (at least) two reasons. First, the level of abstraction has been raised: developers don't have to define their own tree-like structures as libraries exist that can be leveraged. Second, the student body of the Software Engineering Programme now reflects the healthy heterogeneity that is the workforce in software engineering, security, and related industries." [33]

Various aspects of the Programme have been written about previously [6,33,34]. In addition, in [25], the authors considered the relationship between relational database design and the language of Z and explored how the relationship between the two paradigms is exploited within the teaching of the Programme. Finally, the use of a model-driven approach to support the Programme's information system (amongst others) is described in [5].

3.2   Courses and Assignments

To gain a Postgraduate Certificate, a student needs to attend and submit an assignment for four courses (averaging at least 50% across all assignments, with no more than one scoring below 45%); to gain a Postgraduate Diploma, attendance and subsequent submission for eight courses is required (averaging at least 50% across all assignments, with no more than two scoring below 45%); for an MSc, the requirement is 10 (averaging at least 50% across all assignments, with no more than two scoring below 45%), together with the successful completion of a dissertation.

Each course consists of: a period of preparatory study (involving, for example, the reading of one or more research papers or book chapters and/or a small exercise); an intensive teaching week; and a written assignment. Each teaching week involves some combination of lectures and exercise/practical sessions. The relatively small class sizes of up to 18 students typically lead to much interaction between students and instructors. The take-home assignments—which are undertaken over a period of six weeks—allow students to reflect upon and apply the techniques taught during the week.

There are good reasons for this choice of mode of assessment. First, our students often travel from all over the world to attend our courses; to expect them to travel back to sit examinations would be impractical. More importantly, a six-week period in which to undertake an assignment provides students with an opportunity to properly reflect upon the material that was delivered during the one-week course.

3.3 The Software Engineering Mathematics Course

The course under consideration—*Software Engineering Mathematics*—attempts to do two things. First, it attempts to teach students key aspects of discrete mathematics—with the mathematical language of Z being the vehicle of delivery. Second, it aims to show how formal models can be used to aid comprehension and communication.

A 'light touch approach' (as per, for example, the philosophy of [15]) is advocated[3], and a realistic view of the success of the impact of formal methods in general (as reflected by, for example, [24,30]) is presented. The course text is *Using Z* [41] by Woodcock and Davies; *Discrete Mathematics By Example* [32] is used as a supplementary text for additional examples and exercises.

Anecdotal evidence suggests that the course is seen as 'difficult' by many students: the combination of new concepts and techniques, an unfamiliar language, and the intense pace of a week-long course makes for a challenging experience for some students. In addition, this course is a particular victim of the disconnect between theory and practice—while the techniques taught (thinking abstractly and precisely) are clearly beneficial in the long term, this is not always immediately obvious to the students.

Students gain a passing grade in this subject (50%+) if they can demonstrate that they can use the mathematical language of Z to build simple models; they

---

[3] See [43] for a useful classification of 'lightweight formal methods'. While Z does not appear in the discussion, we would argue that it's ideally suited to be used as a 'lightweight' method.

gain a grade in the distinction range (70%+) if they can demonstrate that they can convincingly undertake deductive and inductive proofs.

4   The Approach

We now give consideration to the changes in our approach to delivering the Software Engineering Mathematics course.

4.1   The Motivation for Change

As discussed in Sect. 3, the course text is *Using Z* by Woodcock and Davies [41]. Prior to the change in emphasis, the course's timetable followed faithfully the first 10 chapters of the book:

1. Introduction (Monday AM)
2. Propositional logic (Monday AM)
3. Predicate logic (Monday AM–Tuesday AM)
4. Equality and definite description (Tuesday PM)
5. Sets (Wednesday AM)
6. Definitions (Wednesday PM)
7. Relations (Thursday AM)
8. Functions (Thursday PM)
9. Sequences (Thursday PM–Friday AM)
10. Free types (Friday AM).

The timetable (and, relatedly, the textbook) gave rise to two main challenges in delivering the content. First, natural deduction is at the forefront of *Using Z*: natural deduction rules for conjunction, disjunction, etc. are presented at the point at which the core logical concepts are introduced. To some students, this presents a barrier to learning as the pace at which they learn the notions of propositional and predicate logic is slowed due to a need to appreciate the intricacies of natural deduction rules (and tactics). An additional consequence is that natural deduction assumes greater importance in the minds of the students than it perhaps deserves.

Second, while the timetable allowed for exercises that reinforced learning of individual concepts, there wasn't the scope to allow students to leverage the techniques taught to actually build models: exercises simply reinforced the concepts taught in the previous hour or two. Subsequently, there was evidence that, when it came to the assignment, some students—having not had the experience of building models during the week—had difficulty making the transition from theory to practice.

4.2 A Change in Emphasis

The substantial change made was to compress and redistribute material to ensure that all of the material required to utilise the taught techniques in a meaningful way and build models was taught by the end of Wednesday—leaving Thursday clear for a whole day of case studies. (Friday morning was thereafter dedicated to free types and structural induction).

The other important change (although less important in the context of this paper) was to divorce the introduction to propositional and predicate logic from the introduction to natural deduction. The resulting compressed timetable looked as follows:

1. Introduction and propositional logic (Monday AM): Chapters 1 and 2 (minus natural deduction)
2. Predicate logic, equality and definite description (Monday PM): Chapter 3 (minus natural deduction) and Chapter 4
3. Natural deduction (Tuesday AM): the remainder of Chapters 2 and 3
4. Sets and definitions (Tuesday PM–Wednesday AM): Chapters 5 and 6
5. Relations (Wednesday AM–Wednesday PM): Chapter 7
6. Functions and sequences (Wednesday PM): Chapters 8 and 9 (minus structural induction on sequences)
7. Modelling case studies (Thursday AM and PM)
8. Free types and structural induction (Friday AM): the remainder of Chapter 9 and Chapter 10.

4.3 Benefits and Challenges

The change gave rise to two significant benefits. First, the new structure has a clear delineation between modelling and proof: proof techniques no longer 'get in the way' when introducing new techniques. Second, the conceptually familiar topic of sets appear significantly earlier in the week: on Tuesday afternoon, rather than on Wednesday morning.

As well as benefits, the change gave rise to some challenges. The most significant challenge was that, in order to create the space to spend a whole day on modelling exercises, the pace of the first three days necessarily had to be swift in order to cover the material. Second, as there was deviation from the 'natural order' of the course text, there had to be a degree of trust from the students that the 'postponed' material would be covered in due course.

4.4 An Example

Dedicating a whole day to modelling case studies allows students to apply the techniques that they have been taught. The students tackle the exercises in groups of three or four, using whiteboards. If time allows, the students utilise LATEX and the Fuzz type-checker.

An example case study is reproduced below.

A TV recording system records television programmes to a hard-drive. The hard-drive has the capacity to store up to 200 h of programming; each programme may be at most 6 h in length.

When the viewer accesses the hard-drive, they are presented with a menu presenting all of the shows currently stored. The details are:

- title;
- programme length; and
- whether or not the programme has been viewed.

You may assume the following types and abbreviations.

$$[Title]$$
$$Length == \mathbb{N}$$
$$Viewed ::= yes \mid no$$

(We shall assume that the length of recordings is represented in terms of **minutes**.)

(a) Complete the following axiomatic definition with appropriate constraint information ("the hard-drive has the capacity to store up to 200 h of programming; each programme may be at most 6 h in length"):

$$hd : \text{seq}\,(\,Title \times Length \times Viewed\,)$$

$$\vdots$$

The sequence hd captures information pertaining to programmes stored on the hard-drive. (You should assume, for now, the existence of a function $cumulative\_total \in \text{seq}\,(\,Title \times Length \times Viewed\,) \rightarrow Length$. This function will be defined in part (d).)

(b) Define, via set comprehension, the collection of titles of programmes (which appear in hd) that are over two hours in length.

(c) Define functions $viewed$ and not_viewed that take sequences of type $Title \times Length \times Viewed$, and return the sequences with the not viewed and viewed programmes removed respectively. So,

$$viewed\,\langle (t_1, 3, yes), (t_2, 4, yes), (t_1, 5, no)\rangle =$$
$$\langle (t_1, 3, yes), (t_2, 4, yes)\rangle$$
$$not\_viewed\,\langle (t_1, 3, yes), (t_2, 4, yes), (t_1, 5, no)\rangle =$$
$$\langle (t_1, 5, no)\rangle$$

(d) Define a **recursive** function, $cumulative\_total$, that takes sequences of type $Title \times Length \times Viewed$ and returns the cumulative length of the programmes recorded. So,

$$cumulative\_total\,\langle (t_1, 3, yes), (t_2, 4, yes), (t_1, 5, no)\rangle = 12$$

(e) Give a μ-expression for the title of the longest programme that appears in hd.

(f) Define a function that maps programme titles (which appear in hd) to cumulative lengths, i.e.,

$$f \langle (t_1, 3, yes), (t_2, 4, yes), (t_1, 5, no) \rangle = \{t_1 \mapsto 8, t_2 \mapsto 4\}$$

(g) Define a function that takes a sequence of programmes and removes the longest viewed one, i.e.,

$$g \langle (t_1, 3, yes), (t_2, 4, yes), (t_1, 5, no) \rangle = \langle (t_1, 3, yes), (t_1, 5, no) \rangle$$

(h) Define a function that takes an element of seq ($Title \times Length \times Viewed$) and sorts that sequence in terms of programme length—with the longest programme appearing first. So,

$$s \langle (t_1, 3, yes), (t_2, 4, yes), (t_1, 5, no) \rangle = \\ \langle (t_1, 5, no), (t_2, 4, yes), (t_1, 3, yes) \rangle$$

Typically, such an exercise—which would take an expert no more than 20 min or so to complete—will take groups of three or four between two and three hours.

5   Indicative Results

We now consider some indicative results regarding the success of the initiative. We recognise that there are caveats here: the class sizes are small; the groups are heterogeneous in their make up; there is an element of subjectivity in any assessment process. However, we are able to leverage data spanning several years.

There have been 10 instances of the course using the approach described in this paper; to compare, we also consider the final 10 instances of the course using the former approach. We consider first student performance (in terms of examination results) and then consider student feedback.

5.1   Student Performance

As already discussed, students are assessed by way of a take-home assignment that they have six weeks to complete.

The assessment criteria for the course are given below.

1. Propositional and predicate logic: have you understood the syntax and semantics of propositional and predicate logic? can you write logical statements? can you interpret logical statements? can you reason about logical statements?
2. Equality and definitions: do you understand the notion of a type? do you understand the different ways of introducing types, sets, and identifiers into a formal document? do you understand the notion of equality and its associated properties?

3. Sets, relations, functions, and sequences: do you understand the formal representations of these structures? can you define such structures according to some property? can you apply the operators associated with these structures? can you interpret a statement defined in terms of these operators? can you reason about such statements? can you use these structures to describe systems and properties?
4. Free types: can you define a free type? have you understood the principle of recursion? can you construct an inductive proof?

Table 1. Student performance

Date	Submissions	Min.	Max.	Mean	Median	% 50+	% 70+
2010 (iteration 1)	16	10	85	59	62.5	81.25	37.5
2010 (iteration 2)	13	35	95	61	55	84.6	23.1
2011 (iteration 1)	17	20	85	56	55	64.7	29.4
2011 (iteration 2)	12	35	80	58	57.5	83.3	16.7
2012 (iteration 1)	15	30	85	64	60	86.7	40
2012 (iteration 2)	10	20	90	61	60	90.0	30.0
2012 (iteration 3)	10	30	80	55	56.5	60.0	30.0
2013 (iteration 1)	13	20	90	63	65	84.6	38.5
2013 (iteration 2)	11	50	80	62	58	100.0	36.4
2014 (iteration 1)	13	33	94	65	60	92.3	15.4
2014 (iteration 2)	14	40	95	67	67	78.6	42.9
2015 (iteration 1)	3	55	63	60	63	100.0	0.0
2015 (iteration 2)	19	10	88	61	64	89.5	31.6
2015 (iteration 3)	16	42	80	63	66.5	81.3	43.8
2016 (iteration 1)	5	35	74	55	62	60.0	20.0
2016 (iteration 2)	15	40	88	67	67	86.7	46.7
2017 (iteration 1)	12	35	83	63	65	83.3	41.7
2017 (iteration 2)	16	35	73	58	61	81.3	12.5
2018 (iteration 1)	14	45	95	72	69	92.9	50.0
2018 (iteration 2)	14	55	90	72	71.5	100.0	50.0
Pre-change	**130**	**10**	**95**	**60**	**60**	**82.3**	**35.4**
Post-change	**128**	**10**	**95**	**64**	**65**	**85.9**	**37.5**

Assignments in this subject typically consist of 10 questions and follow a similar structure each time: Question 1 is typically concerned with truth tables; Question 2 is typically concerned with equivalence proofs; Question 3 typically pertains to proof trees; Questions 4–8 leverage a scenario, asking the students to write definitions and constraints, and then define sets, relations and functions. Such questions are on a par (in terms of style and difficulty) with the case studies discussed in Sect. 4. Questions 9 and 10 typically involve free type definitions and proof by induction. The structure of the assignments was consistent across the course instances considered in this paper.

Recall from Sect. 3 that students pass an assignment in this subject (scoring 50%+) if they can use the mathematical language of Z to build simple models; they gain a grade in the distinction range (70%) if they can demonstrate that they can convincingly undertake deductive and inductive proofs.

Table 1 illustrates examination results for 20 iterations of the course: the first 10 were delivered 'pre-change'; the last 10 were delivered 'post-change'. The number of submissions, lowest score, highest score, mean score and median score are given for each instance. The percentage of submissions scoring 50+ and 70+ respectively are also given.

The bottom rows aggregate the respective scores. Curiously, the lowest and highest grades do not differ at all. However, the mean and median scores have improved significantly; there are slight increases in the percentages scoring 50+ and 70+.

There is one final measure that can be utilised: non-submission of assignments by students who have attended the course. This rate has decreased slightly: from 21.2% (pre-change) to 20.0% (post-change).

There is, beyond the raw facts, little that we can conclude here. However, the overall increase in grades is clearly a pleasing result and perhaps indicates that, even if practice does not 'make perfect', it does 'make better'.

Table 2. Student feedback

	All courses	Pre-change	Post-change	**Difference**
Statement 1	4.6	4.73	4.77	*0.85%*
Statement 2	4.66	4.68	4.82	*2.99%*
Statement 3	4.41	4.38	4.59	*4.79%*
Statement 4	4.77	4.83	4.94	*2.28%*
Statement 5	4.46	4.64	4.7	*1.29%*
Statement 6	4.77	4.89	4.91	*0.41%*
Statement 7	4.75	4.84	4.85	*0.21%*
Statement 8	4.55	4.58	4.75	*3.71%*
Statement 9	4.43	4.13	4.18	*1.21%*
Statement 10	4.42	4.45	4.64	*4.27%*
Statement 11	4.44	4.38	4.61	*5.25%*
Statement 12	4.58	4.56	4.7	*3.07%*
Overall	**4.57**	**4.59**	**4.71**	**2.61%**

5.2 Feedback

Following each course, students are invited to complete (anonymously) a questionnaire. The statements (scored in the range 1–5) are as follows.

1. The lectures added significant value to the course material
2. The lecturer took the time needed to explain the key concepts
3. The lectures included valuable contributions from the other students in the class
4. The lecturer was helpful and ready to answer questions
5. The exercises helped me to understand the topics covered in the lectures
6. The lecturer or tutor was knowledgeable and encouraging
7. Help was available—from the lecturer or tutor—when I needed it
8. Issues raised were adequately addressed—through model solutions or discussion
9. I think that the techniques taught during the course will be valuable to me in the future
10. The course was well constructed: the various components worked well together
11. The course material was appropriate, and of good quality
12. The course administration was efficient and effective.

While the final question is not of direct relevance to this paper, we include it here for the sake of completeness.

In Table 2, we compare the scores per-question for pre-change and post-change iterations. We also compare the scores with the overall scores across all courses between mid-February 2010 (when data was first collected in this fashion) and mid-February 2019—giving rise to a total of 6264 completed questionnaires.

When comparing pre-change and post-change courses, there is a positive difference in feedback in all questions. The most significant differences can be seen for Statements 3 ("The lectures included valuable contributions from the other students in the class"), 8 ("Issues raised were adequately addressed—through model solutions or discussion"), 10 ("The course was well constructed: the various components worked well together") and 11 ("The course material was appropriate, and of good quality").

Post-change, the course outperforms the average feedback with respect to all statements, with one exception. (Pre-change, it was below the average on three others: "The lectures included valuable contributions from the other students in the class", "The course material was appropriate, and of good quality" and "The course administration was efficient and effective".) And it is this question—"I think that the techniques taught during the course will be valuable to me in the future"—which, after all, motivated the changes (and, indeed, this contribution). While the slight increase is pleasing, the feedback does, perhaps, show that there is still some way to go in terms of demonstrating relevance to practitioners.

6   Conclusions

In this paper we have described how we have changed the emphasis of a course that introduces part-time students typically employed in the software engineering industry to introductory topics from discrete mathematics and formal methods.

While our arguments for such an emphasis are not new (see, for example, [2], in which Barr advocates helping the situation by requiring students to model real-world implementations), we have been able to demonstrate how, via close to a decade's worth of data, the changes have given rise to some pleasing results.

We recognise that our experiences are somewhat unique: the nature of the Software Engineering Programme (being targeted at professional software engineers) is very different to an undergraduate programme in Computer Science; the make-up of the class is more heterogeneous; the nature of the teaching (in intensive one-week blocks) is different from the typical mode of delivery; and the nature of assessment differs from what most full-time students will be used to. However, we do think that some of the challenges faced will be familiar to many, and, indeed, are part of the ongoing discourse with respect to the value and applicability of modelling techniques.

One clear trend over the 25+ years of the Programme's existence is the shift from companies funding their employees' professional development to few employers now being prepared to provide such support. As it is now typical for students to 'pay their own way', there is an increasing need to provide evidence of practical value. To this end, and reflecting upon the results of Sect. 5, future changes will be driven by the statement "I think that the techniques taught during the course will be valuable to me in the future".

Acknowledgements. The author would like to thank the anonymous reviewers for their helpful and constructive comments.

References

1. Almi, N.E.A.M., Rahman, N.A., Purusothaman, D., Sulaiman, S.: Software engineering education: The gap between industry's requirements and graduates' readiness. In: Proceedings of the IEEE Symposium on Computers and Informatics (ISCI 2011), pp. 542–547 (2011)
2. Barr, T.: Improving software engineering education by modeling real-world implementations. In: Proceedings of the 8th edition of the Educators' Symposium (EduSymp 2012), pp. 36–39. ACM (2012)
3. Cowling, A.J.: The role of modelling in teaching formal methods for software engineering. In: Bollin, A., Margaria, T., Perseil, I. (eds.) Proceedings of the 1st Workshop on Formal Methods in Software Engineering Education and Training (FMSEE&T 2015). CEUR Workshop Proceedings, vol. 1385 (2015)
4. Cristiá, M.: Why, how and what should be taught about formal methods? In: Bollin, A., Margaria, T., Perseil, I. (eds.) Proceedings of the 1st Workshop on Formal Methods in Software Engineering Education and Training (FMSEE&T 2015). CEUR Workshop Proceedings, vol. 1385 (2015)
5. Davies, J.W.M., Gibbons, J., Welch, J., Crichton, E.: Model-driven engineering of information systems: 10 years and 1000 versions. Sci. Comput. Program. **89**, 88–104 (2014)
6. Davies, J., Simpson, A., Martin, A.: Teaching formal methods in context. In: Dean, C.N., Boute, R.T. (eds.) TFM 2004. LNCS, vol. 3294, pp. 185–202. Springer, Heidelberg (2004). https://doi.org/10.1007/978-3-540-30472-2_12

7. Devlin, K.: Why universities require computer science students to take math. Commun. ACM **46**(9), 37–39 (2003)

8. Dijkstra, E.W.: The humble programmer. Commun. ACM **15**(10), 859–866 (1972)

9. Fincher, S., Utting, I.: Pedagogical patterns: their place in the genre. In: Caspersen, M.E., Joyce, D.T., Goelman, D., Utting, I. (eds.) Proceedings of the 7th Annual SIGCSE Conference on Innovation and Technology in Computer Science Education, (ITiCSE 2002), pp. 199–202. ACM, June 2002

10. Finkelstein, A.: Software engineering education: a place in the sun? In: Proceedings of the 16th International Conference on Software Engineering (ICSE 1994), pp. 358–359. IEEE Computer Society Press (1994)

11. Fraser, S., et al.: Meeting the challenge of software engineering education for working professionals in the 21st century. In: Proceedings of the 18th Annual SIGPLAN Conference on Object-Oriented Programming, Systems, Languages, and Applications (OOPSLA 2003), pp. 262–264 (2003)

12. Hayes, I.J.: Specification Case Studies, 2nd edn. Prentice-Hall, Hertfordshire (1992)

13. Honiden, S., Tahara, Y., Yoshioka, N., Taguchi, K., Washizaki, H.: Top SE: educating superarchitects who can apply software engineering tools to practical development in Japan. In: Proceedings of the 29th International Conference on Software Engineering (ICSE 2007), pp. 708–718. IEEE Computer Society Press (2007)

14. Ishikawa, F., Taguchi, K., Yoshioka, N., Honiden, S.: What top-level software engineers tackle after learning formal methods: experiences from the top SE project. In: Gibbons, J., Oliveira, J.N. (eds.) TFM 2009. LNCS, vol. 5846, pp. 57–71. Springer, Heidelberg (2009). https://doi.org/10.1007/978-3-642-04912-5_5

15. Jackson, D.: Lightweight formal methods. In: Oliveira, J.N., Zave, P. (eds.) FME 2001. LNCS, vol. 2021, pp. 1–1. Springer, Heidelberg (2001). https://doi.org/10.1007/3-540-45251-6_1

16. Jackson, M.: Aspects of abstraction in software development. Softw. Syst. Model. **11**(4), 495–511 (2012)

17. Jacky, J.: The Way of Z: Practical Programming With Formal Methods. Cambridge University Press, Cambridge (1997)

18. Jaume, M., Laurent, T.: Teaching formal methods and discrete mathematics. In: Dubois, C., Giannakopoulou, D., Méry (eds.) Proceedings of the 1st Workshop on Formal Integrated Development Environment (F-IDE 2014), pp. 30–43 (2014)

19. Kiniry, J.R., Zimmerman, D.M.: Secret ninja formal methods. In: Cuellar, J., Maibaum, T., Sere, K. (eds.) FM 2008. LNCS, vol. 5014, pp. 214–228. Springer, Heidelberg (2008). https://doi.org/10.1007/978-3-540-68237-0_16

20. Kramer, J.: Is abstraction the key to computing? Commun. ACM **50**(4), 36–42 (2007)

21. Kramer, J.: Abstraction and modelling—a complementary partnership. In: Czarnecki, K., Ober, I., Bruel, J.-M., Uhl, A., Völter, M. (eds.) MODELS 2008. LNCS, vol. 5301, pp. 158–158. Springer, Heidelberg (2008). https://doi.org/10.1007/978-3-540-87875-9_11

22. Kramer, J., Hazzan, O.: The role of abstraction in software engineering. ACM SIGSOFT Softw. Eng. Not. **31**(6), 38–39 (2006)

23. Larsen, P.G., Fitzgerald, J., Riddle, S.: Learning by doing: practical courses in lightweight formal methods using VDM+. Technical Report CS-TR-992, University of Newcastle upon Tyne (2006)

24. Mandrioli, D.: On the heroism of really pursuing formal methods: title inspired by Dijkstra's "On the Cruelty of Really Teaching Computing Science". In: Proceedings of the Third FME Workshop on Formal Methods in Software Engineering (Formalise 2015), pp. 1–5. IEEE Computer Society Press (2015)
25. Martin, A.P., Simpson, A.C.: Generalizing the Z schema calculus: database schemas and beyond. Proc. APSEC **2003**, 28–37 (2003)
26. Mead, N.R., Ellis, H.J.C., Moreno, A., MacNeil, P.: Can industry and academia collaborate to meet the need for software engineers? Cutter IT J. **14**(6), 32–39 (2001)
27. Morgan, C.C., Sufrin, B.A.: Specification of the UNIX filing system. IEEE Trans. Software Eng. **10**(2), 128–142 (1984)
28. Nishihara, H., Shinozaki, K., Hayamizu, K., Aoki, T., Taguchi, K., Kumeno, F.: Model checking education for software engineers in Japan. SIGCSE Bulletin **41**(2), 45–50 (2009)
29. Reed, J.N., Sinclair, J.E.: Motivating study of formal methods in the classroom. In: Dean, C.N., Boute, R.T. (eds.) TFM 2004. LNCS, vol. 3294, pp. 32–46. Springer, Heidelberg (2004). https://doi.org/10.1007/978-3-540-30472-2_3
30. Robinson, K.: Embedding formal development in software engineering. In: Dean, C.N., Boute, R.T. (eds.) TFM 2004. LNCS, vol. 3294, pp. 203–213. Springer, Heidelberg (2004). https://doi.org/10.1007/978-3-540-30472-2_13
31. Roscoe, A.W.: Understanding Concurrent Systems. Springer, London (2010). https://doi.org/10.1007/978-1-84882-258-0
32. Simpson, A.C.: Discrete Mathematics by Example. McGraw-Hill, Boston (2002)
33. Simpson, A.C., Martin, A.P., Cremers, C., Flechais, I., Martinovic, I., Rasmussen, K.: Experiences in developing and delivering a programme of part-time education in software and systems security. In: Proceedings of the 37th International Conference on Software Engineering (ICSE 2015), vol. 2. pp. 435–444. IEEE Computer Society Press (2015)
34. Simpson, A.C., Martin, A.P., Gibbons, J., Davies, J.W.M., McKeever, S.W.: On the supervision and assessment of part-time postgraduate software engineering projects. In: Proceedings of the 25th International Conference on Software Engineering (ICSE 2003), pp. 628–633. IEEE Computer Society Press (2003)
35. Spivey, J.M.: The Z Notation: A Reference Manual, 2nd edn. Prentice-Hall International, Englewood Cliffs (1992)
36. Subrahmanyam, G.V.B.: A dynamic framework for software engineering education curriculum to reduce the gap between the software organizations and software educational institutions. In: Proceedings of the 22nd IEEE International Conference on Software Engineering Education and Training (CSEET 2009), pp. 248–254 (2009)
37. Tarkan, S., Sazawal, V.: Chief chefs of Z to alloy: using a kitchen example to teach alloy with Z. In: Gibbons, J., Oliveira, J.N. (eds.) TFM 2009. LNCS, vol. 5846, pp. 72–91. Springer, Heidelberg (2009). https://doi.org/10.1007/978-3-642-04912-5_6
38. Vaughn, R.B., Carver, J.: Position paper: The importance of experience with industry in software engineering education. In: Proceedings of the 19th IEEE International Conference on Software Engineering Education and Training (CSEET 2006), p. 19 (2006)
39. Warford, J.S.: An experience teaching formal methods in discrete mathematics. ACM SIGCSE **27**(3), 60–64 (1995)
40. Wing, J.M.: Computational thinking. Commun. ACM **49**(3), 33–35 (2006)
41. Woodcock, J.C.P., Davies, J.W.M.: Using Z: Specification, Refinement, and Proof. Prentice Hall, Englewood Cliffs (1996)

42. Woodcock, J.C.P., Larsen, P.G., Bicarregui, J., Fitzgerald, J.: Formal methods: Practice and experience. ACM Comput. Surv. **41**(4), Article number 19 (2009)
43. Zamansky, A., Spichkova, M., Rodriguez-Navas, G., Herrmann, P., Blech, J.O.: Towards classification of lightweight formal methods. In: Proceedings of the 13th International Conference on Evaluation of Novel Approaches to Software Engineering (ENASE 2013) (2018)

Teaching Formal Methods: Lessons Learnt from Using Event-B

Néstor Cataño[(⊠)]

Innopolis University, Kazan, Tatarstan, Russia
nestor.catano@gmail.com

Abstract. This paper summarises our experience in teaching Formal Methods to Computer Science and Software Engineers students from various universities around the world, including the University of Madeira in Portugal, the Pontificia Universidad Javeriana and the University of The Andes in Colombia, Carnegie Mellon University (CMU) in the USA, and Innopolis University (INNO) in Russia. We report challenges we have faced during the past 10 to 15 years when teaching formal methods using the EVENT B formalism, and describe how we have evolved the structure of our courses to respond to those challenges. We strive to help students to build skills on Formal Methods that they can employ later on in their future IT jobs in software Industry. Our goal is to promote the wide use of Formal Methods by software Industry. We consider that this goal cannot be achieved without first universities transferring to Industry students with a strong background in Formal Methods and related formal tools. Formal Methods are key to software development because they are based on Discrete Mathematics which can be used to properly reason about properties that the software one develops should have. We have conducted two surveys among our students, the first one at CMU and the second one at INNO, that we use here to document and justify our decisions in terms of the course structure. The first survey is about the use of EVENT B as main mathematical formalism, and the second one is about the organisation of teams of students within the classroom to work on software projects that use EVENT B as main mathematical formalism. Our hope is that our work can be reused by other Faculty to make their own decisions on course structure and content in the teaching of their Formal Methods courses.

Keywords: Computer Science · Discrete Mathematics · EVENT B · Formal Methods · Software Engineering

1 Introduction

This paper summarises our experiences in designing and teaching an EVENT B [1] master course on Formal Methods. The MSS (Models of Software Systems) course is a Computer Science course lectured to Software Engineers students at

© Springer Nature Switzerland AG 2019
B. Dongol et al. (Eds.): FMTea 2019, LNCS 11758, pp. 212–227, 2019.
https://doi.org/10.1007/978-3-030-32441-4_14

Carnegie Mellon University (CMU) in Pittsburgh, USA, and at Innopolis University (INNO), Russia. Decisions made on course structure have been shaped by our own previous experience in teaching Formal Methods from 2004 to 2015, which includes giving various guest lectures to students of the Programming Usable Interfaces (PUI) master course offered at CMU to HCI (Human Computer Interaction) students in 2010. This is particularly important because HCI students do not necessarily have a strong background in Logic and Discrete Mathematics, which poses additional challenges to us. To our knowledge, that was the first time that two, arguably, diverse topics, Formal Methods and HCI, are combined into a single master course to formally develop Android apps.

The Models of Software Systems master course is part of the MSIT-SE (Master of Science in Information Technology-Software Engineering) programme offered at both CMU and INNO. The goal of the programme is to create software company leaders in the field of Software Engineering and to help students build theoretical as well as practical expertise in the use of Formal Methods techniques which they can later use in their careers. Models of Software Systems is a Formal Methods course taught to Software Engineering students in the Fall of every year. It exposes students to several formalisms including first-order logic, state machines, concurrency, and temporal logic. In the Fall of 2016, we commenced to teach at INNO an adapted version of the Models of Software Systems course offered at CMU. This adapted version profited from our previous experience with PUI whereby students implemented a usable and verified Android app during their course project. The adapted version offered at INNO has been nurtured by the results of two surveys. The first survey (Sect. 3) was conducted in Pittsburgh in the Fall of 2015 among students of the Models of Software Systems course offered at CMU. The survey sheds light on the benefits of teaching EVENT B to SE students. The Models of Software Systems course at CMU includes a course project with 3 deliverables for the modelling and analysis of an Infusion Pump [4]. At INNO, we restructured the course project to consider the analysis and formal software development of an Android app. The second survey (Sect. 4) was conducted among students of the Models of Software Systems course at INNO. The goal of this second survey is to understand how students can work together as a team to develop software modelled with EVENT B.

The primary goal of this paper is to outline several aspects and guiding principles that can be used by other Faculty to structure their courses on EVENT B. These aspects and principles take into consideration various features and peculiarities of our students (introduced in Sect. 2). This paper is about teaching Formal Methods and formal software development with EVENT B. Our work does not intend to serve as a reference to teach any other SE subject. In our course, software development with EVENT B does not compete with software development with any other technique, and indeed, we always encourage students to combine Agile methodologies with software development with EVENT B during their course projects.

The rest of this paper is structured as follows. Section 2 presents our characterization of our students. We have formulated this characterization by direct interaction with them and through discussions with other Faculty. Section 3 explains the structure of the Models of Software Systems course that is lectured at INNO and the results of the first survey. This survey led to a list of recommendations to re-structure the Models of Software Systems course which we also discuss. Preliminary results of this first survey were presented to SECM'17 [5]. Section 4 presents the results of the second survey. Section 5 discusses related work and Sect. 6 presents conclusions and discusses future work.

2   Our Students

It is commonly understood that in order to teach well one should know in advance whom one is teaching to. We present our characterisation of our students below. This characterisation should be taken as a heuristic of recommendations of common students traits. It has been gathered through daily interaction with students in and out of the classroom. Of course, each future below is not common to every and each student.

Tech Savvy. Our students are not afraid of technology. They naturally engage in technology and the use of novel devices. They interact with each other to seek information.

Discovery-Driven. Our students are often interested in the most recent technological inventions of society. Students apply novel approaches to today's problems.

Immediate Feedback. Our students quite often ask for immediate feedback about the activities they undertake, or feedback on the results of their assessments. They expect tools would give them immediate feedback. For instance, students expect feedback from proof assistants (called provers) on why a proof rule cannot be applied at a certain point during the proof, or why a proof tactic cannot discharge a whole proof. They see provers as push-button technology. They expect immediate feedback from compilers about which line of code produces a particular error.

Elaborative Rehearsal. We encourage elaborative rehearsal in our courses in which students not just repeat a concept (a definition, a proof, a program, a proof tactic, etc.) over and over again but memorise the technique behind the concept to apply it when it should be.

Our support to the practice of "Elaborative Rehearsal" started very early in our teaching of Formal Methods, back in 2006, when the author was invited to give a couple of guest lectures on JML [6,11] (Java Modeling Language) and Design-by-Contract (DbC) [14] for the final part of an undergraduate course on EVENT B offered at Pontificia Universidad Javeriana (PUJ) in the Spring of every year. We wanted to bridge mathematical models in EVENT B, whose syntax was familiar to students, to Java programs and DbC JML contracts, which was a new topic for them. Therefore, we designed and implemented the EVENTB2JAVA Java code generator [15], which generates JML-specified Java programs for EVENT B models. This constituted a breakthrough in the way we started lecturing EVENT B as the code generator was used to assist in development of course projects ever since.

Active Learner. Our students easily engage in activities they are interested in. They frequently discover strategies through individual experiments with a tool, for instance, when using provers, they might apply pruning steps of automatic proofs and restart the proof with a different proof assistant.

Easily Bored. Our students get bored with things that are not interesting to them. Things that interest them are often related to technology, social activities, media, and the Internet.

Visually Focused. Our students are interested in systems and programs they can picture in their minds. Traditional Formal Methods courses use toy examples to introduce topics and theories. Students are often not interested in or struggle to understand those types of examples. They often prefer to be presented examples they can visualize in their minds, or are related to some particular technology they are familiar with.

Multi-Tasker. Our students are often involved in multiple activities at the same time, which may or not relate to Academia.

Team Focus. Our students often struggle with working in teams.

Socially Aware. Students often engage in social activities. They care about society, animals, nature, and other people around them. They enjoy social media and social apps; they engage in social activities.

Learning from Failure. This is related to Immediate Feedback. Students learn through failure and use counter-examples to validate their theories. Students feel the need for examples that contradict or confirm their theories.

3   The Survey at Carnegie Mellon University

MSS (Models of Software Systems) students at CMU (Carnegie Mellon University) have previous exposure to logic and software development, typically covered by courses such as Discrete Mathematics and Software Engineering. The MSS course at CMU consists of 16 weekly classes and 16 weekly recitation sessions. Sessions are 2 h and 45 min each. The course has homework assignments, which are issued weekly and are due the following week. Each recitation session discusses issues and challenges that took place during the homework assignment of the previous week. Students are exposed to propositional and predicate logic, proof techniques, sets relations and functions, sequences and induction, state machines, Z [17], concurrency, and linear temporal logic.

The following survey encompasses 3 main questions related to EVENT B. The survey was conducted among the students of the MSS course at CMU. It was anonymous and conducted online. 29 students answered it. Answers were not mandatory, so some students left some answers blank. The goal of the survey is to sense students' opinions on the advantages and disadvantages of using EVENT B as mathematical formalism; these opinions are valuable to us students were exposed to various mathematical formalisms prior to learning EVENT B. The questions of the survey are based on two hypotheses that the survey attempts to corroborate or refute. EVENT B has a practical lien to code refinement and code generation that is not quite present in other mathematical formalism.

Hypothesis 1: Students can understand a program written in EVENT B easily.

Hypothesis 2: EVENT B can easily be integrated and used to validate, verify, animate, and reason about software systems used in the industry.

The following are the questions of the survey:

Question 1: Overall, how would you rate the EVENT B sessions?

Answer	#
Excellent	4
Good	18
Neutral	6
Poor	1
Terrible	0

The results for this first question of the survey show that about 76% of the students answered Good or Excellent, 21% answered Neutral, and 3% answered Poor or Terrible.

Question 2: What was your favourite part of the EVENT B sessions?

Answer	#
They were motivated by real examples	9
The close link between EVENT B, code generation, and programming languages	6
EVENT B syntax is easy to understand	6
EVENT B is tool-supported	2
I like it overall	1
Nothing	1
Left blank	4

79% of the answers given to this question point out to practical aspects of EVENT B. By "real examples" students mean a strong connection to software systems. Students were presented with modelling example of a social network in EVENT B [7]. MSS is rather an unusual course. It is a Formal Methods course, and hence strongly mathematically oriented, lectured to SE students, who might or might not be as mathematically strong as Computer Science students often are. This fact compels us (Formal Methods instructors) to motivate and attract students by presenting modelling and verification examples of applications they use in life rather than demonstrating traditional Computer Science toy examples. The examples must illustrate the lien between modelling and verification with software technology.

Question 3: Which aspects of EVENT B did you find attractive or unique (that you do not find in other formalisms)?

Answer	#
Its approach to software development	6
Its support for code generation	6
Its tool support	2
I do not know	1
None	1
Left blank	12
It is easy to use	12

48% of the answers given (the first 3 rows) point out to practical aspects of EVENT B. The first row makes reference to the fact that EVENT B implements refinement calculus techniques.

Question 4. What would make the EVENT B sessions better?

Answer	#
6	More lectures
5	More examples, including code generation demos
2	Putting EVENT B sessions right after Z sessions
16	Left blank

38% of the answers (the first 2 rows) point out to extending the sessions on EVENT B. The third row points out to having those sessions right after the sessions on Z as notations of both languages are similar.

Overall, Question 1 tells us about students' general satisfaction on the EVENT B part of the course. Second and fourth answers to Question 2 provide support for hypothesis 2. The last answer to Question 3 and the third answer to Question 2 give some indications about Hypothesis 1.

Next, we relate answers to the survey with the features presented in Sect. 2. The two first answers to Question 2 relate to "Visually Focused". We presented examples related to Facebook. This further seeks to stress the "Social Aware" feature of our students. The second answer to Question 4 is related to the "Elaborating Meaning" feature of our students. Students always like to see programs running, in particular if those are programs for a logical model. From the results of the survey above, we decided to write a series of recommendations to modify the structure of the MSS course. Some of these modifications have successfully been implemented at INNO. We map each recommendation to one or several features presented in Sect. 2.

3.1 Recommendations

The survey led to the elaboration of some recommendations for improvement of various aspects of the MSS (Models of Software Systems) course. Modifications to the MSS course are subtle due to the tight interplay of the course material, hence, lectures are assessed throughout weekly homework assignments, which are related to the course project, the midterm, and the final exam. Hence, introducing EVENT B into the course syllabus entails to create a homework assignment for each EVENT B session, to add relevant questions on EVENT B to the midterm and final exams, and primarily to link EVENT B to one or all the three project deliverables. Here is our list of recommentations.

1. Build a large set of modelling examples and homework assignments with questions and solutions. Examples must be full-fledged modelling and verification examples of software systems. This suggestion is to be realised by writing the second part of the teacher guide book so as to include modelling and verification examples of software systems.
   Related Aspects: (*i*) "Visually Focused", examples must relate to systems students are familiar with rather than to programs. (*ii*) "Discovery-Driven", examples relate to mobile applications or social networking sites. (*iii*) "Elaborative Meaning", examples expose students to practical aspec first and then the theory is introduced on-the-fly as examples need it.
2. During the recitation sessions, we should conduct logical proofs in Coq rather than attempting pencil-and-paper proofs only.
   Related Aspects: (*i*) "Immediate Feedback", the Coq tool provides immediate feedback on errors users make during a proof. (*ii*) "Learning from Failure", feedback provided by the Coq tool enables users to learn from their mistakes. (*ii*) "Active Learner", Coq enables students to discover when proving strategies will work better, to abandon a strategy when it does not work and then apply another one.
3. Although the previous suggestion is about using Coq, other proof assistants could be used too. Nevertheless, the advantage of using Coq is related to the Curry-Howard isomorphism: a mathematical proof in classical logic without the excluded-middle rule is a program in the logic of the typed lambda-calculus. Students can run proofs as programs, for instance, in Objective Caml (an implementation of typed lambda calculus). If a lecture introduces a soundness proof of the translation performed by a parser, then the proof is

just the program implementing the parser. There is no better way to motivate students to conduct proofs: proofs are programs that are part of software systems students can run.

Related Aspects: (*i*) "Elaborative Meaning", the theory about the syntax and semantics of a parser is carried down to animating a program that shows what the parser does.

4. Regarding the lectures on Natural and Structural induction, the key link between induction and programming is recursion. Recursive definitions require well-founded inductive proofs. If Coq is to be introduced into the MSS course, one can use Objective Caml (Coq's programming language) to write examples of recursive definitions, and Coq to formalize the algorithm in logic and discharge underlying Proof Obligations. Examples of recursive definitions may relate to data structures, for instance, for searching algorithms.

   Related Aspects: (*i*) "Elaborative Meaning", the intrinsic aspects of recursive definition proofs are boiled down to running programs in OCaml.

5. We plan to incorporate the teaching of software development with EVENT B into the MSS course (Check questions of Survey 2 in Sect. 4). EVENT B enables users (*i*) to use a tool (Rodin [2]) to write mathematical models about sets (In EVENT B, relations are sets of pairs), (*ii*) to tool-check whether the set-based model is correct, (*iii*) to conduct correctness proofs about set-based models, (*iv*) and to generate Java code (via the EVENTB2JAVA tool) for students to animate formal models of software systems.

3.2   Implementation of Recommendations at Innopolis University

This section explains which of the previous recommendations have we implemented at Innopolis University (INNO).16 Regarding recommendation 1, we almost finished writing the second part of the guide book.

We have not implemented recommendations 2 and 3 at INNO. Introducing Coq to the course syllabus would require a lot of effort regarding preparation. It would require us (*i*) to re-structure the slides of the first part of the course (about 25%), (*ii*) to re-work the homework assignments to be based on Coq, and (*iii*) to adapt the course project to account for Coq.

We have not implemented recommendation 4. Let us discuss an example of how this recommendation can be implemented in the classroom. We define a Stack in Objective Caml that implements standard operations. We ask students to implement a Map function that takes a function and a Stack object and applies the function to each element of the Stack. The result is a new Stack obtained by applying the function to each element of the original Stack. The Map function can be defined recursively. One would ask students whether their recursive definitions are correct or not, asking them to undertake the correctness proof formally. The Objective Caml program can naturally be re-written in Coq, where the proof can be conducted.

Regarding recommendation 5, we extended the teaching of EVENT B to 4 sessions of two hours 45 min each. Each weekly session has its respective homework assignment on EVENT B. We re-oriented the third deliverable of the course for students to design in EVENT B the core functionality of an Android app [9]. The Android app is structured following an MVC design pattern. The VC part is based on OpenGL, the M part is initially be modelled in EVENT B and then code generated to Java using the EVENTB2JAVA tool. For the third course-project deliverable, students must conduct 4 tasks. The first task asks students to use ProB [12] to detect any likely deadlock or race condition in the EVENT B model. The second task asks students to define safety properties in EVENT B. The third task asks students to generate code, to animate it, and to check if the code runs as expected. The fourth task asks students to re-implement the interface of the app.

4   The Survey at Innopolis University

Formal software development with EVENT B follows "the parachute strategy" in which systems are first considered from a very abstract and simple point of view, with broad fundamental observations. This view usually comprises a few simple invariant properties that students can easily grasp, for instance, defining what can reasonably be expected from the operation of such a system. When writing a model for a software system in EVENT B students should write an abstract *machine* (model) and then successively write *refinement* machines [3]. For each refinement machine Proof Obligations (POs) are to be discharged to ensure that it is a proper refinement of the most abstract machines. Only once all the machines are written and all the POs are discharged one can consider the underlying system has completely been modelled. But, if an abstract machine is modified, for instance, invariants are added to it, or some definition is changed, then typically new POs are generated for all the machines in the refinement chain, or existing proofs are to be re-run. The worst scenario happens when a software requirement changes or a new one is added on top of the existing ones as this typically would break existing invariants. Pedagogically speaking this raises a concern regarding the way members of a software development team should work together and how team members can share their workload. If team members work together in a way that each member is in charge of designing and tool-proving the correctness of a sole machine, then each time a member introduces a change, the work of any team member in charge of a refinement machine will become invalid. In an opposite direction, one team member can be in charge of writing the whole model, but then, at least from a pedagogical perspective, this will diminish the EVENT B learning curve of the other team members. The parachute strategy advocates for the Waterfall software development methodology in which software requirements are set upfront and then the software development process starts. In practice, this is quite difficult to achieve, and even if it is achieved, it is often the case that actual definitions are changed on-the-fly, for instance, when one decides to encode a variable with a total and not with a partial function, invalidating all the related and discharged POs.

The survey presented in Sect. 4.1 relates to issues regarding how a group of students can work together as a team to develop software with EVENT B. The survey was conducted at INNO (Innopolis University) among students of the MSS course, it was anonymous and conducted online. Each of the 25 students answered the survey. The survey seeks to address the "Team Focus" feature described in Sect. 2.

4.1   Student's Feedback

We gathered software requirements for WhatsApp from our experience using it. We focused on WhatsApp's Android mobile version and disregarded its web version. After we wrote the initial software requirements document we proceeded to formalise the requirements in EVENT B. However, it was often the case that we had discussions with students in and out of the classroom to clarify our understanding of the functionality of WhatsApp. For instance, when two persons are chatting and one of them decides to delete a previously sent content (message, picture or video), shall this content be deleted from the sender, the receiver or anyone to whom the content has been forwarded too? Would this functionality (to delete a content item) be implemented differently if the person who is deleting the content is the sender (the person who sent the content initially) or the receiver of the content? All these questions required careful discussions both in and out the classroom as different implementations would break the invariants written for WhatsApp.

The above text gives the reader an introduction to the first and second questions of the survey. We thought that students could follow a Waterfall style of software development, but we finally needed to evolve the software requirements document.

Question 1: What do you think would be the most suitable software development methodology to develop WhatsApp with EVENT B and Rodin?

1. Agile (requirements evolve, change at any time)
2. Waterfall (requirements are stable, don't change)
3. Both combined
4. Other? Which one?

Answer	#	%
Agile	6	24%
Waterfall	9	36%
Both	8	32%
Other	2	8%

By looking at the answers to the first question, students are more or less equally fine with developing WhatsApp following Waterfall, Agile, or combining both methodologies. In the end, students needed to combine both methodologies as requirements changed.

As for the last row of results, 2 students selected Spiral as software development methodology, which goes in the direction of a software project in which software requirements evolve.

Question 2: Did your team develop WhatsApp following the methodology selected in Question 1?

Answer	#	%
Fully	1	4%
Largely	10	40%
Fairly	9	36%
Scarcely	4	16%
Not at all	1	4%

According to the results above, 80% of the students (the 3 first rows) followed a software methodology that they considered the most suitable. We gave students complete freedom so as to choose any software methodology that they considered the most appropriate to develop WhatsApp.

Question 3: If you decide to develop WhatsApp following an MVC design pattern structure, how do you think your team should be organized to develop the M (model) part of WhatsApp?

1. Software requirements are fixed in advance, and each team member develops one or several different machines; the team members meet at an early stage to decide who (which team member) will develop what functionality and which machine; in the end, team meets again to put all the machines together.
2. Only two team members would develop the complete functionality of WhatsApp in EVENT B; the other two or three members would provide continuous feedback to the first two members. In short, you would engage in a "pair programming" discipline of working organized in groups of two members.
3. None.

Answer	#	%
Fixed	6	24%
Paired	18	72%
None	1	4%

Changes in software requirements are particularly cumbersome in software development with EVENT B since they might affect one machine and therefore all the machines in its refinement-chain making often most of the discharged POs invalid afterwards. Students can then decide to split the number of machines (4 for this project) into equal shares among students (4 or 5 students per team), working individually and communicating decisions regularly as a team, or they can select some of the team members to work in the EVENT B formalization and the rest of the members to work, for instance, in Android, in the visual interface of the app. But then, students were also concerned about learning EVENT B properly as this was included in the final exam. 72% of the students selected the last option of team work (second row in the results table) in accordance with an Agile methodology of work in which requirements change constantly.

Question 4: How difficult was for you to use EVENT B to model the M (Model) part of WhatsApp?

Answer	#	%
Very hard	6	24%
Hard	12	48%
Moderate	7	28%
Easy	0	0%
Very easy	0	0%

72% of the students (the first 2 rows) found difficult to come up with an implementation of the core functionality of WhatsApp. The initial difficulty was to write a sound model for WhatsApp in EVENT B. Additional difficulties came from the use of the EVENTB2JAVA tool which did not support some of the syntax of EVENT Bso that students needed to manually write the code generated by the tool in some cases.

Question 5: How difficult was for you to extend the code generated for the M (model) part of WhatsApp so that it can be used from the V (view) part?

Answer	#	%
Very hard	5	20%
Hard	13	52%
Moderate	4	16%
Easy	3	12%
Very easy	0	0%

Students needed to extend the core functionality of WhatsApp in the following way. They needed (i) to define the architecture of their implementation of WhatsApp, and (ii) either implement it or use an existing platform that could handle concurrency of several users chatting with each other in several chat-rooms. Students needed to write some wrapping code that links the interface of the app developed with Android Studio with the code generated by EVENTB2JAVA for the core functionality of the app. 72% of the students (the first 2 rows) considered that implementing this extension was difficult, which was expected by course instructors.

Question 6: Given flexible time and project conditions, which approach would you use to bridge/interface the Java code generated for the M part of WhatsApp to the implementation of its V part?

1. You would write EVENT B code for the extended functionality of the M part functionality and would generate code to Java with the EVENTB2JAVA tool that interfaces with the V part of WhatsApp.
2. You would manually and directly implement the extended functionality in Java that interfaces with the V part of WhatsApp.
3. Both combined.
4. None.

Option	#	%
Code generation	1	4%
Manual implementation	13	52%
Both combined	4	40%
None	1	4%

Question 6 seeks to sense students' opinion about using EVENT B for developing the interface of WhatsApp. 52% of the students though it is not worthwhile to use EVENT B for that purpose, only 4% of them thought it is, and 40% of them thought they could attempt a combined effort. We considered that it would be preferable to write the interface manually given the complexity and size of graphical libraries of Android.

4.2 Related Aspects

At INNO, the course project was changed in accordance with the Related Aspects discussed in Sect. 2.

Tech Savvy. Though most of our students have prior experience in programming, only a bunch of them have prior experience in programming with the Android platform. Hence, working on an Android project during the MSS course gave students the opportunity to learn a new technology while working on mathematical formalisms behind the scenes.

Immediate Feedback. In our courses, EVENT B is introduced with the aid of the Rodin IDE [2], a platform that provides support for writing models in EVENT B. Rodin comes with a series of provers that give students feedback when discharging POs (Proof Obligations).

Visually Focused. During the third project-deliverable students implement a visual interface of the Android app that links to its core functionality.

Socially Aware. Examples of social apps introduced in our courses include a social event planner (an app to invite people to gather around a social event), WhatsApp, among others, all of which can be framed as course projects.

5   Related Work

In [13], Méry presents a teaching programming methodology using EVENT B. We do not teach program development but system development in our courses. Nevertheless, program development can be tackled as a last program refinement step in the EVENT B formalisation of our course projects, and EVENTB2JAVA can be extended to generate Java program implementations.

Early efforts in introducing Formal Methods have been made in the past. In [18], Stanley Warford describes his efforts to incorporate Formal Methods in the teaching of Discrete Mathematics. The author argues that Formal Methods can be mastered at the undergraduate level. He claims that "the benefit of teaching Formal Methods in a Discrete Math course had immediate benefit in following courses". Jaume and Laurent [10] share the same view that Formal Methods should be taught during the undergraduate studies. We also think that our master course would benefit from students having previous exposure to Formal Methods, for instance, during their undergraduate studies. We found that our students find easier to understand concepts such as injective, surjective or bijective functions through the use examples related social networks or systems related to technology.

More recently [16], Ken Robinson describes his experience in teaching Formal Methods. He claims that "Software Engineers should aspire to fault-free software", which goes in the same direction as our teaching. We believe in developing correct software from specifications to code.

In a similar manner to our work, Gibson and Méry [8] report on lessons learnt on the teaching of Formal Methods. They adopt a similar view to ours in the sense of expecting that Formal Methods students can be transferred from Academia to Industry and expect that transferred students can make Formal Methods more popular in Industry. Authors work in small case studies in an intuitive manner so that students can appreciate the need for formality.

6   Conclusion

Organising a course project around software development teams helps students to enhance their collaboration spirit. Our students are always motivated about any course project that involves the use of Android as they are always attracted to technology. The use of the EVENTB2JAVA code generator to implement the course project is one of our assets. Students are always positively surprised to see how mathematical models based on predicate calculus relate to programs written in Java (or another programming language). They love to execute mathematical models to get a grasp on their behaviour.

We would like to mention one of the difficulties that our students have regarding Learning from Failure. Students love to learn from failure and the use counter-examples to check if something is right or wrong, but the logical meaning they attach to counter-examples is often wrong. If one says "most water bottles are made of plastic", then students might think the sentence is not true because they know "a water bottle made of glass", without realizing that the two sentences do not conflict each other. The second sentence does not make the first sentence invalid. To help students understand the first sentence one would need to add some redundancy, let us say, "most but not all the water bottles are made of plastic".

Our future work is mainly related to completing the guide book with modelling examples and homework assignments as described in Sect. 3.1. Each example includes (*i*) the core functionality of the example written in EVENT B, (*ii*) all the POs discharged with Rodin, (*iii*) an implementation of the functionality generated with the EVENTB2JAVA tool, and (*iv*) an interface implementation, for instance, an Android app implementation with Android studio. We have also started writing a second book, more oriented to Java practitioners, to help them effectively use EVENT B to enhance the quality of their programs. The book does not go much into details about discrete mathematics and first-order logic but introduces only the mathematic formalism needed by the book to explain the Java programming examples.

References

1. Abrial, J.-R.: Modeling in Event-B: System and Software Design. Cambridge University Press, New York (2010)
2. Abrial, J.-R., Butler, M., Hallerstede, S., Hoang, T.S., Mehta, F., Voisin, L.: Rodin: an open toolset for modelling and reasoning in Event-B. Softw. Tools Technol. Transf. **12**(6), 447–466 (2010)
3. Abrial, J.-R., Hallerstede, S.: Refinement, decomposition, and instantiation of discrete models: application to Event-B. Fundamentae Informatica **77**(1–2), 1–28 (2007)
4. Arney, D.E., Jones, P., Lee, I., Ray, A.: Generic infusion pump hazard analysis and safety requirements version 1.0. Technical report, University of Pennsylvania (2009)
5. Cataño, N.: An empirical study on teaching formal methods to millennials. In: First International Workshop on Software Engineering Curricula for Millennials (SECM/ICSE), Buenos Aires, Argentina. ACM and IEEE Digital Libraries (2017)
6. Cataño, N., Barraza, F., García, D., Ortega, P., Rueda, C.: A case study in JML-assisted software development. In: Machado, P., Andrade, A., Duran, A. (eds.) Brazilian Symposium on Formal Methods (SBMF). Electronic Notes in Theoretical Computer Science, vol. 240, pp. 5–21, July 2009
7. Catano, N., Rueda, C.: Matelas: a predicate calculus common formal definition for social networking. In: Frappier, M., Glässer, U., Khurshid, S., Laleau, R., Reeves, S. (eds.) ABZ 2010. LNCS, vol. 5977, pp. 259–272. Springer, Heidelberg (2010). https://doi.org/10.1007/978-3-642-11811-1_20

8. Gibson, J.P., Méry, D.: Teaching formal methods: lessons to learn. In: 2nd Irish Workshop on Formal Methods, Cork, Ireland, 2–3 July 1998 (1998)
9. Google Inc.: The Android Platform (2017). http://developer.android.com/design/index.html
10. Jaume, M., Laurent, T.: Teaching formal methods and discrete mathematics. In: Proceedings 1st Workshop on Formal Integrated Development Environment, F-IDE 2014, Grenoble, France, 6 April 2014, pp. 30–43 (2014)
11. Leavens, G.T., Baker, A.L., Ruby, C.: Preliminary design of JML: a behavioral interface specification language for Java. ACM SIGSOFT **31**(3), 1–38 (2006)
12. Leuschel, M., Butler, M.: ProB: a model checker for B. In: Araki, K., Gnesi, S., Mandrioli, D. (eds.) FME 2003. LNCS, vol. 2805, pp. 855–874. Springer, Heidelberg (2003). https://doi.org/10.1007/978-3-540-45236-2_46
13. Méry, D.: Teaching programming methodology using Event B. In: Habrias, H. (ed.) The B Method: from Research to Teaching, Nantes, France, July 2008. Henri Habrias, APCB
14. Meyer, B.: Applying "design by contract". Computer **25**(10), 40–51 (1992)
15. Rivera, V., Cataño, N., Wahls, T., Rueda, C.: Code generation for Event-B. Int. J. Softw. Tools Technol. Transf. (STTT) **19**, 1–22 (2015)
16. Robinson, K.: Reflections on the Teaching of System Modelling and Design. AVOCS (2010)
17. Spivey, M.: An introduction to Z and formal specifications. Softw. Eng. J. **4**(1), 40–50 (1989)
18. Stanley Warford, J.: An experience teaching formal methods in discrete mathematics. SIGCSE Bull. **27**(3), 60–64 (1995)

You Already Used Formal Methods
but Did Not Know It

Giampaolo Bella[✉]

Dipartimento di Matematica e Informatica, Università di Catania, Catania, Italy
giamp@dmi.unict.it

Abstract. Formal methods are vast and varied. This paper reports the essentials of what I have observed and learned by teaching the Inductive Method for security protocol analysis for nearly twenty years. My *general finding* is something I realised after just a couple of years, that my target audience of post-graduate students with generally little appreciation of theory would need something different from digging deep down in the wonders of proof ever since class two. The *core finding* is a decalogue of steps forming the teaching methodology that I have been developing ever since the general finding became clear. For example, due to the nature of the Inductive Method, an important step is to convey the power and simplicity of mathematical induction, and this does not turn out too hard upon the sole basis that students are familiar with the informal analysis of security protocols. But the first and foremost step is to convince the learners that they already somewhat used formal methods, although for other applications, for example in the domains of Physics and Mathematics. The argument will convey as few technicalities as possible, in an attempt to promote the general message that formal methods are not extraterrestrial even for students who are not theorists. This paper introduces all steps individually and justifies them towards the general success of the teaching experience.

1 Introduction

Formal methods form a very big chapter in the book of, at least, Informatics. It is widely recognised that they include a variety of approaches, for example, logic, algebraic or ad hoc approaches. With a "universal view" of formal methods, I contend that hey have been applied to virtually every real-world problem areas, ranging from Astrophysics to Economics to Engineering.

It is clear that my view of a formal method is broad, in fact I like to include in the pool any mathematically grounded, rigorous method. The distinctive feature implied here is that formal methods do not necessarily require the target phenomenon or system under study to be practically available or built at all. As opposed to empirical methods, formal methods can be profitably used on paper, ideally with some computer support, namely at the abstract, design level.

My main "local" preconditions are that students are not very inclined to theory in general. Broadly speaking, I find course modules more geared to practical

© Springer Nature Switzerland AG 2019
B. Dongol et al. (Eds.): FMTea 2019, LNCS 11758, pp. 228–243, 2019.
https://doi.org/10.1007/978-3-030-32441-4_15

competences such as (imperative) programming and system administration. In the cybersecurity area in particular, the most job-oriented competences lie in the area of Vulnerability Assessment and Penetration Testing (which I also introduce at Master's level), hence formal methods again suffer this particular though well motivated trend of the present time. However, also formal methods continue to contribute to the goodness of cybersecurity [1], for example as it can be read from recent publications such as a NIST survey [2] or an NSF workshop report [3]. Hence, the motivation for this paper.

The teaching experience within such a large area as formal methods is bound to be diverse and multifaceted, and here I only engage into outlining my own, limited experience on teaching the Inductive Method [4,5], which is embedded in the theorem prover Isabelle [6], for the analysis of security protocols. This would be the first encounter of my students with theorem proving and formal methods in general. A fundamental disclaimer stemming from my local preconditions is that none of my observations should be taken as general; by contrast, they are limited to the specific lecturing experience in my Institution, though over nearly two decades, at Master's level in Informatics covering a module of at least 12 h intertwining theory and laboratory experiments tightly.

My general finding is that the entanglements of proof theory must be left to an advanced module, which I have never had the opportunity to teach. My core finding is that teaching an introductory module requires at least a decalogue of steps before any proof can be attempted profitably. Lecturing will resemble the tailor's activity of sewing together patches of different fabrics, though dealing with somewhat heterogeneous notions from Informatics in our case.

To try and speed up readability, the style I take in this paper will be mixed, sometimes describing the steps of the decalogue and summarising parts of the lectures, sometimes as if I were speaking straight to the students. Hopefully, the context will resolve the inherent ambiguity. As we shall see, the main obstacle to overcome for students will be the perception of formal methods as something so theoretical and abstract to be unattractive and unsurmountable, hence the title of this paper. But the decalogue discussed below has yielded a very effective teaching experience with me.

2 My Experience with Teaching the IM

There is no room for introducing the Inductive Method [4] and the theorem prover Isabelle [6], so I must assume a basic familiarity of the readers'.

2.1 You Already Used Formal Methods

One of the first issues I encountered since the beginning and that I keep touching every year is some sort mental resistance that (my) students show to almost anything prefixed with "formal". In consequence, there seems to be some psychological wall between themselves and formal methods in general.

At first, I set out to try and demolish that wall upfront. I started providing vast reference material, also appealing to books that can be found freely on the Internet [7], and presenting example applications to various scenarios in the areas of both hardware and software. However, this did not work, as the class felt kind of lost through the various methods, with each student looping through a contrastive analysis of the methods.

I decided that this approach was too vast. So, I selected First-Order Logics and tried to illustrate and variously demonstrate why it could be somewhat easy to use in practice, and also nice and ultimately rewarding; but all this did not seem to yield the results I was expecting. It was clear that students were almost memorizing notions and formulas rather than adopting and actively using them.

It was still in the early years of teaching when I started to feel that psychological wall as unsurmountable for them. So, I thought that the only way to have students on the other side would have been to make this true by assumption. I was then left with the problem of finding an appropriate, realistic interpretation that would make that assumption hold, which would have made students feel already beyond the wall. At some point, I thought I found that interpretation, and presented them with something as simple as this formula:

$$s = v \cdot t$$

This was the first encouraging result because everyone could recognise the *uniform linear motion* formula with s indicating space, v for speed and t for time.

I decided to navigate this way and this is when I decided to take a somewhat loose definition of formal methods. So, I claimed that formula to be an application of a formal method, precisely a *specification*, namely some sort of abstract representation of a real-world phenomenon. The formula clearly shows independence from the actual phenomenon, it lives and computes in a world of its own, that of symbols with a clear, non-ambiguous interpretation. Yet, the formula models and describes the phenomenon closely, providing a realistic, written representation of it. I did not need to describe the language of the formula more in detail because I realised that students had already started to stair at the board pensively, so they were finally engaged.

I then unfolded the same argument with accelerated linear motion and projectile motion. Then, I switched application area, and discussed definite integrals as a very useful tool (not just to pass A levels but also to) calculate the area under a curve, something that we could effectively use to help a farmer determine the extension of his land. Formal methods everywhere! Yes, such formulas are formal because they leave (almost) no room for ambiguity but they are also very applied due to what they allow us to do and resolve in everyday life. This argument worked with the class, definitely, and keeps working every year.

I normally conclude this journey through heterogeneous applications of formal methods with an extra reference to Propositional Logics and First-Order Logics, whose basics the students regularly know from some foundational course. This time around, they look at whatever I try to formalise with these languages with renowned interest and, as far as I can tell, more familiarity and conscious understanding. For example, here I normally debate that there is no "wrong" specification of a phenomenon but, rather, there may be an unrealistic specification of it, for example like describing a river that flows from sea to peaks.

"Dear student, it is clear that you already used formal methods but did not know it!".

2.2   The Need for Formal Methods and in Particular for the IM

The next step in the decalogue is to demonstrate that formal methods are needed in general. Here, it is useful to go back to the formula borrowed from Physics and Calculus, as well as to hint at Ancient Greece mathematician Eratosthenes, with his incredibly precise measurement of Earth's circumference, and other Ancient Greece prodigies.

To approach our days, I normally linger around the Pentium processor bug (which luckily has a Wikipedia entry). With whatever microchip in hand to test, it is intuitive for students to see in their minds the act of feeding it with various inputs to inspect whether the output is correct. And here are the fundamentals of modern (industrial strength) testing. However, the Pentium bug shouted out to the world that testing may not be enough. This may be due, in general, to the ever increasing complexity of modern circuits, whose complexity roughly doubles every 18 months, as Moore started to predict ever since 1965. It may also be due to the tight time-to-market constraints of products, and this is likely to have been the case with the Pentium bug. While it is natural for everyone that testing requires time due to the number of tests to physically execute, the learners also understand that business success often correlates with early deployment. (An underlying, usefully embodied, assumption is that even if something is appropriately designed, it is not obvious that it will work as expected at design level when it is actually built, such as with houses or with any devices).

And here comes a clear need for an alternative that scientists may use, on paper or arguably with some computer assistance, to get confidence that the real-world phenomenon that is an actual industrial product works as intended by its designers. That alternative is the use of formal methods, whose application may not be constrained by execution times as testing is. This argument invites at least two useful considerations. One is that formal methods support some sort of *reasoning* on the target phenomenon, formal reasoning in fact, which can be tailored to assess specific properties of interest, (functional ones) such as correctness of computation, then (non-functional ones such as) secrecy and authentication. Another useful argument is the predictive use of formal methods. We can effectively study a phenomenon before it actually takes place, or a product before it is built, and this is an exclusive advantage.

This is the point when it becomes effective and useful to plunge into security protocols, thus nearing my actual target. Students are normally familiar with traditional attacks on toy security protocols, which are so popular in the literature of the area. For example, I use to entertain my undergraduates with an *informal* analysis of the original public-key Needham-Schröder protocol [8], and I always succeed in convincing everyone that nonces remain secret and that mutual authentication works; after that, I surprise them with Lowe's attack, then help them overcome their frustration by observing that the attack was only published some 17 years after the protocol. Therefore, I easily emphasise the limitations of informal protocol analysis, calling for more rigour, hence for formal protocol analysis, which has the strength and rigour of mathematics. Examples are due here, but they still need to wait one more logical assertion.

That assertion is that a security protocol may be a strange, huge beast. There is potentially no bound for the length of protocol messages, for the number of protocol steps, of protocol participants, of nonces or keys they may invent and for the number of protocol sessions they may interleave. It becomes apparent that security protocols are potentially unbounded in size, hence it becomes intuitive that the empirical approach of testing (all) its potential executions falters. Consequently, the idea that some sort of mathematical wisdom could help starts to materialise at the mental horizons of the students.

Additionally, familiarity with the Needham-Schröder protocol implies acquaintance with the notion of threat model, and in particular with the standard Dolev-Yao attacker. Because that attacker may intercept messages and build new ones *at will* with the sole limitations imposed by encryption, students realise that the attacker is yet another source of potential unboundedness, and know by intuition that modelling it may not be straightforward.

Even if we took the approach of bounding all parameters and we magically knew that the resulting protocol reached its security goals, then we still would have no guarantee that those goals hold also when those parameters are exceeded during a practical use. It would seem that unboundedness cannot be neglected.

"So, dear all, you will be amazed at how the Inductive Method can cope with unboundedness!".

2.3   A Parallel: How to Write a Biography

At this point, some students change the way the look at the lecturer, as if they start to wonder independently how to possibly use the Inductive Method to model security protocols. Here, I surprise them turning to talk about biographies, actual people's biographies. The biographer faces a huge challenge: to condense a continuous (we could build a bijection with the reals) sequence of events in a finite manuscript. The biographer has no option except picking up a few significant events and describe those, perhaps connecting them logically, and sometimes drawing a general message about the chief character, either explicitly or implicitly. From a data structure standpoint, a biography is a list of events.

The same approach can be taken to model security protocols, somewhat surprisingly for students. So, our effort could be similar to the biographer's. Running a protocol of course entails a number of tasks for each of its peers. But, as the biographer does, we need to *abstract away* from many of those and distill out the main ones. With security protocols, it is easy to convince everyone that the main ones are to send and to receive a cryptographic message.

Does this imply that a protocol can be compared to a human life? It would seem so in terms of modelling effort and approach. More precisely, a specific protocol execution can be compared to the biography of a life, and both can be modelled as a list of events. While a biography features events linking the chief character to other people in the character's life, or sometimes other people among themselves, a protocol execution features events linking peers to each other via the events of sending or receiving the protocol messages (a specific peer could be isolated and interpreted as a chief character in the execution but this is irrelevant). A list representing a protocol execution is normally termed a *trace*, hence it is a list of events of sending or receiving the protocol messages. We could then address a biography as the trace of someone's life. If we blur the focus on the chief character, then a biography is a representation of one possible development of society, simply because it may involve many characters.

This argument invites thinking about other possible executions of protocols in parallel to other possible developments of society, the very society of people on this planet. And here students find themselves curious to understand if and how all of the protocol executions and, equally of the society developments, could be represented compactly. They get all the more hungry as they start to perceive that such possible executions or developments are potentially unbounded. They will have to resist their hunger a bit longer.

"We now know that a list is a useful structure to model an abstract version of one possible protocol execution, but how can we ever model all possible executions?".

2.4   The Use of Computational Logics for Reasoning

At this point in the development of the discourse, the learners' eyes begin to glitter. It is hence the right time to instill the power of logics. I already mentioned that, in my experience, students normally come with some notions of Propositional Logics and First-Order Logics, and discussed how to make them feel familiar with such logics (Sect. 2.1). However, it would seem that logics is merely seen as a language for *specifying* (or *formalising*) some phenomenon. It is then not very clear to them what to do with a specification or how to use it profitably. Here come handy again the arguments unfolded above, suggesting that a specification is a somewhat compact representation of something real (Sect. 2.3) and that it may be used to understand that thing predictively (Sect. 2.2).

The only way to overcome the dogmatic flavour that such justifications may bring is to finally introduce elementary forms of classical reasoning to be conducted on top of specifications, with the aim of *proving* something about the specification. My favourite one is modus ponens, so I normally draw something like this on the board:

$$\frac{p \to q \quad p}{q}$$

Stating that *"if you have $p \to q$ and you also have p, then you may also derive q"* is simply not enough to convey the meaning of this essential rule.

Students have often taken $p \rightarrow q$ alone to magically derive q. This betrays their misunderstanding, whereas $p \rightarrow q$ and p are both preconditions at the same logical level, and it is precisely their combination what allows us to derive q; here, it may help to denote p as the ammo that the weapon $p \rightarrow q$ needs to shoot out q.

This is an essential yet powerful form of (formal) reasoning, and it may also be the students' first close encounter with such a wonderful engine that, once they have certain formulas that hold, allows them to derive yet another formula that holds too. Should the learners show concern that they are touching something extraterrestrial again, I easily demolish that concern asserting that we all follow an essential rule: if I am hungry, then I eat something. At every moment in time, each of us is left wondering: am I hungry? It is clear that, only when this is affirmative, do both preconditions of the modus ponens rule hold, hence it is time to eat something. We all use modus ponens in all sorts of ways.

"Guys, you have only scratched the surface of formal reasoning, still you shall see that you'll be able to do a lot with what you just found out!".

2.5 The Basics of Functional Programming

Functional programming is, for some reasons beyond the aims of this paper, not very well received by my students, who tend to see it again as something overly formal and not as actual programming. Convincing them fully of the power of functional programming normally remains out of reach despite the fact that they took a short crash course (which, however, lacks the details of Turing completeness). The main issue that I take pains to convey is that it is just a *different* programming paradigm from their dearest imperative approach, the latter learned since school. They find it bewildering that a functional program has no variables to assign values to.

So, how on earth can we carry out any sort of computation? The notion of term rewriting must be introduced. Each rewriting derives from the application of a sound rewriting rule. For example, linking the argument back to the use of logics for reasoning (Sect. 2.4), modus ponens may be seen as a rewriting rule for the pair of facts forming its preconditions. Similarly, $p \rightarrow q$ can be rewritten as $\neg p \lor q$ by applying the logical equivalence of the two formulas as a rule.

But rewriting may also be conditional. For example, evaluating the guard of:

$$X = (\text{if } 2 + 1 = 3 \text{ then } Y \text{ else } Z)$$

allows us to rewrite the entire expression as $X = Y$. And this was computation.

"Rewriting is the essence of computation with functional programming, stop thinking imperatively here, forget variables and assignments!".

2.6 The Wonders of Mathematical Induction

Students are somewhat familiar with mathematical induction, in particular for what concerns the definition of the natural numbers:

Base. $0 \in \mathbb{N}$
Ind. if $n \in \mathbb{N}$ then $suc(n) \in \mathbb{N}$

Because they understand rule Ind, it is the right time to introduce its more formal version:

$$\text{Ind. } n \in \mathbb{N} \implies suc(n) \in \mathbb{N}$$

This lets me motivate the meta-level implication as the implication at the level of reasoning, as opposed to the object level of the encoded logics. And I can then introduce an equally formalised version of modus ponens:

$$[\![p \longrightarrow q; \; p]\!] \implies p$$

I believe that with this and a few similar examples, the level of reasoning, as expressed by fat square braces, semicolon and the fat arrow, is uploaded.

And here is how beautiful it is to capture a clearly unbounded set by means of just two, formal, rules. Observe also the magic behind induction, at least due to the fact that nobody has ever tried to practically verify if, say, 4893 can be effectively built by an application of rule Base and a finite number of applications of rule Ind. Still, we know by intuition that all natural numbers are represented.

Observing that *all* natural numbers are caught this way brings back memories of an open problem, how to capture all possible society developments or protocol executions. The answer clearly is *by induction* but we need to cope with traces. Traces are lists, so can we build lists by induction? Yes, we can build them by structural induction on their length. Therefore, we expect to be able to specify all possible lists, even if unbounded, for our application, be is society or protocols, with just a few inductive rules. And, of course, yes, we may have more than one inductive rule in an inductive definition.

Before giving an example (Sect. 2.7), it is useful to go back to the reasoning part and observe that it may also follow predefined *strategies* aimed at proving a goal, thus *proof strategies*. Induction may also be viewed as a proof strategy, based on application of the *mathematical induction proof principle*. As it is a principle, there is no proof for itself. I often realise that students are able to prove facts such as an expression for the sum of the first n natural numbers S_n:

$$S_n = \frac{n \cdot (n+1)}{2}$$

They mechanically prove it for the Base case and then for the Ind case; in the latter, they know how to assume the property, say, for n and then attempt to prove it on that assumption for $n+1$. They may, however, not be fully aware that they are inherently applying the induction proof principle. It is then important to spell it out formally on a property P:

$$[\![P(0); \; P(n) \implies P(n+1)]\!] \implies \forall n. \; P(n)$$

I have memories of their surprise in front of this formal statement. This version is also useful to teach that the latest occurrence of n is scoped by the universal quantifier, hence not to be confused with the occurrences in the preconditions.

"You see now, induction is great for specifying and then for reasoning, namely for formalising something and then proving facts about it!".

2.7   The Formal Protocol Definition

All pieces of the puzzle are now available to compose a formal protocol model. As noted above, students are familiar with toy security protocols at least, hence there will be little to discuss in front of an example like this:

$$1.\ A \longrightarrow B : A, N_a$$
$$2.\ B \longrightarrow A : \{N_a, B\}_{K_b^{-1}}$$

Initiator A sends her identity along with a fresh nonce of hers to responder B, who sends it back, bundled with his identity, encrypted under his private key. The formal model for this example protocol is finally unveiled as shown in Fig. 1. I normally spend above an hour explaining it. It must be first looked at "from the outside-in", namely you must first realise the general structure of what you have in front. It is five rules. They very often mention `fep`. This is a constant (not a variable!) termed as an acronym for *f*formal *e*xample *p*rotocol, which is the formal model for our example protocol. I purposely defer the discussion of its type till now. Because we wanted to formalise all possible protocol executions, and each execution was a trace, namely a list of events (of sending or receiving the protocol messages), then `fep` must be a set of lists of events.

```
Base:   "[] ∈ fep"

Fake:   "⟦evsf ∈ fep; X ∈ synth(analz(knows Spy evsf))⟧
        ⇒ Says Spy B X # evsf ∈ fep"

Fep1:   "⟦evs1 ∈ fep; Nonce Na ∉ used evs1⟧
        ⇒ Says A B {Agent A, Nonce Na} # evs1 ∈ fep"

Fep2:   "⟦evs2 ∈ fep; Gets B {Agent A, Nonce Na} ∈ set evs2⟧
        ⇒ Says B A (Crypt (priSK B) {Nonce Na, Agent B}) # evs2 ∈ fep"

Recp:   "⟦evsr ∈ fep; Says A B X ∈ set evsr⟧
        ⇒ Gets B X # evsr ∈ fep"
```

Fig. 1. Definition of `fep`, the formal model for our example protocol

Going back to the rules defining `fep`, it can be seen that the first rule is very special because it has no preconditions. It is in fact the base case of the inductive definition. While the central rules, `Fep1` and `Fep2`, seem to be "similar" to the protocol steps, the final rule, `Recp`, seems to be a matter of receiving messages, but must be explained in depth. The remaining rule, `Fake`, is incomprehensible without close inspection.

It must be noted that all rules following `Base` mention a trace of `fep` in the preconditions, respectively *efsf*, *evs1*, *evs2*, *evs3* and *evsr*. Recalling that # is the list cons operator, it may also be seen that all those rules conclude that the respective trace, somehow extended on the left, is a trace of *fep*. These features signify that they are all inductive rules. So, we are facing a definition with a total of four inductive rules.

Rule *Fep1* models the first step of the protocol. Standing on a trace *evs1* of the model, it also assumes a nonce that is not used on the trace, hence the nonce is fresh. Of course, *used* is a function that is defined somewhere, but its definition can wait till later (Sect. 2.8). The nonce freshness is a requirement set by the protocol designers, hence we as analysers are merely adding it to our specification. Event *Says A B ⦃Agent A, Nonce Na⦄* is a self explaining formalisation of the first event of the protocol, and also its justification through the datatype of events can wait (Sect. 2.8). The postcondition of the rule states that the given trace *evs1*, appropriately extended with the event that models the first protocol step, is a trace in the model. Thus, the rule's structure resembles that of Ind (Sect. 2.6).

Rule *Fep2* models the second step of the protocol. It rests on a trace *evs2* with the special requirement that it features an event formalising *B*'s reception of the first protocol message, *Gets B ⦃Agent A, Nonce Na⦄*, and *set* casts a list to a set. The rule concludes that the suitably extended trace is in the model.

The difference between these two rules is that *Fep1* puts no requirement on the trace in terms of traffic occurred on it, so the rule may fire at any time, modelling the real-world circumstance of any agent who may initiate the protocol at any time and with any peer. By contrast, *Fep2* may only fire upon a trace that has already recorded the reception of the first message of the protocol.

If the first message is sent through *Fep1*, what makes sure that it is received? The first message, and in fact *any* message that is sent, is received through rule *Recp*. It insists on a trace on which a generic agent *A* sends a generic message *X* to a generic agent *B*, and extends it with the event whereby *B* receives *X*.

We are left with the *Fake* rule, which models the attacker, arguably represented as *Spy*. The trace extension mechanism is clear, so it can be seen that this rule extends a given trace with an event whereby the attacker sends a fake message *X* to a generic agent *B*. The fake message is derived from a set modelled as a nesting of three functions, from the inside-out, *knows*, then *analz*, finally *synth*, which are to be explained separately (Sect. 2.8). Intuitively, such a nesting simulates all possible malicious activity that a Dolev-Yao attacker can perform, yet without any cryptanalysis.

Thus, the formal protocol model features a number of rules that equals the number of steps in the protocol, augmented with three extra rules, one for the base of the induction, one for the attacker and another one to enable message reception. Thanks to the wonders of induction, set *fep* will have all possible traces that can be built under the given protocol, thus modelling effectively all possible protocol executions. For example, it contains a trace on which ten agents begin the protocol with other agents but none of those messages is received, a trace that sees an agent sending off a message to another one and that message being received many times. We are guaranteed that all possible traces that can be built by any interleaving of the given rules appear in `fep`. It is now time to declare that the specific font indicates that the formal protocol model can be fed, as is, to Isabelle, which will parse it and ensure at least type coherency.

"*And here is how we ultimately define the formal protocol model in Isabelle, including all possible protocol executions under the Dolev-Yao attacker!*".

2.8   The Main Functions

Intuitively, the innermost set, `Knows Spy evsf` contains all messages that are ever sent on `evsf` by anyone. Then, function `analz` breaks down all messages of the set into components, for example by detaching concatenated messages and by decrypting cypher-texts built under available keys (no cryptanalysis at all). Finally, `synth` uses available components to build messages by means of concatenation or encryption, still under available keys.

Suppose that each event in the trace `evsf` carried not a cryptographic message but some... bread, a ciabatta. Then, `knows Spy evsf` would be a set of ciabattas. Suppose that ciabattas are one week old, hence too hard to be eaten. We could then decide to grind them off finely into powder, and this is captured by set `analz(knows Spy evs)`. If we want to mix this strange kind of flour again to try and build bread again (ignoring other ingredients), then the resulting fresh ciabattas would all be in the set `synth(analz(knows Spy evs))`.

The definitions of such functions, of `used` and of the relevant types have been published in many other places [4,5], but I want to stress the didactic value of the definition of `knows` hence report it in Fig. 2. After justifying the declaration, the focus turns to the primitive recursive style, with two rules. Rule `knows_Nil` describes the knowledge that a generic agent `A` can form on observing an empty trace. It reduces to the agent's initial knowledge, formalised as `initState A`, but it must be remarked that "state" is used loosely here and, in particular, it bears no relation to the states underlying model checking.

The other rule pertains to a generic trace, and separates the case in which agent `A`, whose knowledge is being defined, is the attacker from the case in which she is not. For each of these, the definition emphasises the latest event `ev` in the trace, which is then split up as trace `ev # evs`. It can be seen that knowledge is evaluated accordingly to the specific event, which can be the sending of a message, the reception of a message or a third type. This third type was introduced by Paulson with the work on TLS of 1999 [9]. He needed to enable agents to somewhat record the Master Secret of that protocol, and decided that defining an additional event for agents' notes was a convenient way.

I then take a good amount of time to describe why and how the definition makes sure that the attacker knows everything that is sent by anyone or noted by compromised agents, those in the set `bad`. This is the students' first encounter with such a set, and I will surprise them later showing that the set is only declared but never ever defined: all reasoning that follows will be typically parameterised over such a set. It means that the attacker has a full view of the network traffic. Incidentally, the attacker does not need to learn the messages that are received because these must have been sent in the first place, a theorem that can be proved thanks to the definition of rule *Recp*, already discussed. I need also time to explain that any agent who is not the attacker only learns from messages that she sends, receives or notes down herself, because, by being honest, she only has a limited view of the network traffic.

```
consts
 knows   :: "agent ⇒ event list ⇒ msg set"
primrec
  knows_Nil:   "knows A [] = initState A"
  knows_Cons:  "knows A (ev # evs) =
   (if A = Spy then
     (case ev of
        Says A' B X ⇒ insert X (knows Spy evs)
      | Notes A' X  ⇒ if A'∈bad then insert X (knows Spy evs)
                                  else knows Spy evs
      | Gets A' X   ⇒ knows A evs)
    else
     (case ev of
        Says A' B X ⇒
              if A=A' then insert X (knows A evs) else knows A evs
      | Notes A' X  ⇒
              if A=A' then insert X (knows A evs) else knows A evs
      | Gets A' X   ⇒
              if A=A' then insert X (knows A evs) else knows A evs))"
```

Fig. 2. Definition of function *knows*

However, no matter how long I spend to signify this definition, students will be left thirsty for some form of computation. Tight in the mental shackles of imperative programming, they may still strive to see this definition as a rewriting rule that will, by its application, drive and determine computation. A few examples are due. Expression *knows Spy (Says A B X # evs)* will get rewritten, by application of rule *knows_Cons*, as *insert X (knows Spy evs)*. I sometimes need to stress that the resulting expression is simpler because *knows* is applied to a shorter trace; and, once more, that this rewriting is computation.

"You see, this is the core of the Inductive Method in Isabelle, one rule to capture Dolev-Yao, a bunch of rules for the entire formal protocol model!".

2.9   The Basic Interaction with the Theorem Prover

With all instruments on the workbench, it is time to discuss how they can be used practically in Isabelle. It must be noted that those instruments only form the essentials of the Inductive Method, and that the full suite can be found by downloading Isabelle, then inside the \src\HOL\Auth folder. Precisely, all constituents of the Inductive Method are neatly divided into three theory files: *Message.thy*, *Event.thy* and *Public.thy*. While the first two theory names are intuitive, the last perhaps is not fully so. In fact, it originally only contained an axiomatisation of asymmetric, or public-key cryptography, while symmetric, shared-key cryptography was in a separate file `Shared.thy`. However, `Public.thy` now contains both versions, and the other file has been disposed with.

Students are now ready to download Isabelle, find these theory files and familiarise with their contents. Depending on the available time, I may parse the specification part of all three theories. Before continuing to the proof part, I must introduce the fundamental proof methods, which can be applied by means of Isabelle command **apply**, and outline what they do:

- `simp` calls the *simplifier*, namely the tool that applies term rewriting meaningfully; for example, to operate the rewriting just discussed (Sect. 2.8), the analyst needs to call **apply** (`simp add:knows_Cons`)
- `clarify` performs the obvious steps of a proof, such as applying the theorem that deduces both p and q from $p \land q$;
- `blast` launches the classical reasoner, and the analyst may easily state extra available lemmas for the reasoner to invoke;
- `force` combines the simplifier with the classical reasoner;
- `auto` is similar to `force` but, contrarily to all other methods, applies to the entire *proof state*, namely to all subgoals to prove.

I purposely keep the discussion on the proof methods brief because I aim at providing the students with something they can fire and see the outcome of. This will favour their empirical assessment of the proof as it unfolds. Of course, each method is very worth of a much deeper presentation, but this can be deferred depending on the aim of the course module and the available time.

"And now you have commands to try and see marvellous forms of computation unfolding through a proof!".

2.10   Proof Attempts

And finally comes the time to show students how the instruments just learned can be used in practice over an example security protocol chosen from those that have been treated in the Inductive Method. A good choice could be to pick the theory for the original public-key Needham-Schröder protocol, theory `NS_Public_Bad.thy`, which also shows how to capture Lowe's attack.

I open the file and review the formal protocol model for the protocol. I continue arguing that one of the main protocol goals is *confidentiality* and debate how to capture it in the Inductive Method. If we aim at confidentiality of a nonce N, we would like the attacker to be unable to deduce it from her malicious analysis of the observation of the traffic. It means that we leverage a generic trace, then apply `knows` and finally `analz`, and we would aim at `Nonce N` $\notin$ `analz(knows Spy evs)`. After skipping on various lemmas in the file, I reach the confidentiality theorem for the initiator's nonce `NA`, quoted in Fig. 3, and there is a lot to discuss: the preconditions of a trace of the protocol model `ns_public` that features the first protocol message, so as to pinpoint the nonce whose confidentiality is to be proved, NA; the two involved peers assumed not to be compromised; spies to be interpreted as a translation for knows Spy (due to backward compatibility: Paulson originally defined spies [4], then I generalised it as knows [5]).

```
theorem Spy_not_see_Na:
  "[Says A B (Crypt(pubEK B) {Nonce NA, Agent A}) ∈ set evs;
   A ∉ bad; B ∉ bad; evs ∈ ns_public]
   ⟹ Nonce NA ∉ analz (spies evs)"
apply (erule rev_mp)
apply (erule ns_public.induct, simp_all, spy_analz)
apply (blast dest:unique_NA intro: no_nonce_NS1_NS2)+
done
```

Fig. 3. Confidentiality of the initiator's nonce in NS_Public_Bad.thy

The main effort must be devoted to playing with and understanding the proof script. The first proof method that is applied resolves the goal with rev_mp, hence I introduce it as an implementation of modus ponens with swapped preconditions, then sketch the basics of resolution on the fly. Of course, I mostly leverage the intuition behind. Suppose you want to get rich and you are so lucky as to find a secret recipe that guarantees you that whatever you want to reach, you just need to do a couple of things to reach it. What would you then do? You would engage to accomplish that very couple of things. The same reasoning is implemented here through the first command, which leaves us with the preconditions of rev_mp left to prove.

I then dissect the second command as a condensed syntax to apply three proof methods. The first resolves the only subgoal currently in the proof state with the inductive proof principle that Isabelle instantiates on the inductive protocol definition. Isabelle builds it automatically and makes it available as a lemma on top of any inductive definition; it is ns_public.induct in this case.

Here comes the general meta strategy that, after induction, we normally apply simplification, namely term rewriting, and then classical reasoning. This justifies simp_all, which solves the Base subgoal. And we are left facing the Fake case, which Paulson designed a special method to solve, spy_analz. It may be safely applied as a black box for the time being, but it can be dissected, if time, by following another article of mine [10].

The next part of the lecture explains the two lemmas that are applied by blast, discussing the general differences between a destruction rule and an introduction rule, and understanding that the + symbol reiterates the same command on all subgoals. It is didactic to assess which subgoal really requires application of which lemma, so that students also familiarise with forward-style reasoning.

Finally, the same argument is repeated on the confidentiality conjecture on the responder's nonce NB. The proof attempt for this conjecture, omitted here for brevity, cannot be closed, and we are left with a subgoal that describes Lowe's attack whereby the attacker learns NB. It is normally illuminating to note how the prover suggests, actually teaches us, scenarios that are so peculiar that we may not have known them by intuition. Every time such a subgoal remains, and we decode that the reasoning cannot be taken forward, then either we need to change line of reasoning entirely or we understand that the scenario indicates an attack (the very attack that contradicts the conjecture).

3   General Lessons Learned and Conclusions

Formal methods are great help over innumerable application scenarios, and the Inductive Method remains a very effective tool that may at least serve exploratory reasoning on new systems or security goals, possibly to inspire the subsequent implementation of ad hoc tools.

In particular, Paulson also formalised the notion of an *Oops* event ever since the inception of the Inductive Method to allow and agent to arbitrarily lose a secret to the attacker, without any particular precondition. The socio-technical understanding of cybersecurity and privacy is a very hot area today, grounding non-functional properties not only on technical systems such as security protocols but also on the use that humans may make of them. I believe that the *Oops* event is the unique ancestor of all recent works in this area.

The problem treated in this paper was how to transmit the above messages to post-graduate students with an embodied preconception that they do not like theory. While it may be obvious that the contents must be taught gently and incrementally, what I find less obvious is to convince them that they already somewhat used formal methods although they did not use to call them so.

Another far from obvious finding I distilled over the years towards teaching this discipline is the critical review, brought through the creases of my decalogue, of some useful notions they may already have. For example, induction, or just its basics, must be understood profoundly. And the essence of functional programming must be leveraged for the students' proof experience to near their embodied imperative programming experience. I myself insisted on teaching them.

References

1. Parnas, D.L.: Really rethinking 'formal methods'. Computer **43**, 28–34 (2010)
2. Schaffer, K.B., Voas, J.M.: Whatever happened to formal methods for security? Computer **49**, 70 (2016)
3. Chong, S., et al.: Report on the NSF workshop on formal methods for security (2016)
4. Paulson, L.C.: The inductive approach to verifying cryptographic protocols. IOS J. Comput. Secur. **6**, 85–128 (1998)
5. Bella, G.: Formal Correctness of Security Protocols. Information Security and Cryptography. Springer, Heidelberg (2007). https://doi.org/10.1007/978-3-540-68136-6
6. Wenzel, M.: The Isabelle/Isar reference manual (2011). http://isabelle.in.tum.de/doc/isar-ref.pdf
7. MISC: Formal Methods (2019). https://www.freetechbooks.com/formal-methods-f28.html

8. Boyd, C., Mathuria, A.: Protocols for Authentication and Key Establishment. Information Security and Cryptography. Springer, Heidelberg (2003). https://doi.org/10.1007/978-3-662-09527-0
9. Paulson, L.C.: Inductive analysis of the internet protocol TLS. ACM Trans. Comput. Syst. Secur. **2**, 332–351 (1999)
10. Bella, G.: Inductive study of confidentiality, for everyone. Form. Asp. Comput. **26**, 3–36 (2014)

Author Index